VOLUME 1
CHAPTERS 1–8

INTERACTIVE ORGANIZER

INTERACTIVE ALGEBRA FOUNDATIONS

Elayn Martin-Gay
University of New Orleans

Pearson

17 2022

ISBN-13: 978-0-13-471916-0
ISBN-10: 0-13-471916-6

Table of Contents

*Answers are available to instructors in the companion MyLab Math course instructor resource section; instructors can make answers available to students at their discretion.

Section 1.1 Introduction to the Interactive Assignment

Objectives
Objectives   A Get Oriented to the Interactive Assignment

Directions: Complete your Interactive Organizer by filling in the blanks and solving exercises as you complete each screen of the Interactive Assignment.

- For **WORK WITH ME** exercises, follow along and write each step needed and shown to solve, including the final answer.
- For **YOUR TURN** exercises, write the exercise generated for you in MyLab Math, then "show your work" by writing each step needed to solve, including the final answer.

Objective A: Get Oriented to the Interactive Assignment

Watch the objective video.

YOUR TURN #1:   **YOUR TURN #2:**

1

Section 1.2 Place Value, Names for Numbers, and Reading Tables

Objectives
A Find the Place Value of a Digit in a Whole Number
B Write a Whole Number in Words and in Standard Form
C Write a Whole Number in Expanded Form
D Read Tables

Directions: Complete your Interactive Organizer by filling in the blanks and solving exercises as you complete each screen of the Interactive Assignment.
- For **WORK WITH ME** exercises, follow along and write each step needed and shown to solve, including the final answer.
- For **YOUR TURN** exercises, write the exercise generated for you in MyLab Math, then "show your work" by writing each step needed to solve, including the final answer.

Objective A: Find the Place Value of a Digit in a Whole Number

The position of each digit in a number determines its _____ .

WORK WITH ME #1.

Determine the place value of the digit 5 in each whole number.
 a. 657   b. 5423

YOUR TURN #1:   **YOUR TURN #2:**

Objective B: Write a Whole Number in Words and in Standard Form

Watch the objective video.

   Each group of three digits is called a _____ .

Writing a Whole Number in Words
To write a whole number in words, write the _____ in each period followed by the _____ of the period. (The ones period is usually not written.) This same procedure can be used to _____ a whole number.

VIDEO WORK WITH ME.
Write in words.
26,990

Writing a Whole Number in Standard Form
To write a whole number in _____ form, write the number in each period followed by a _____ .

VIDEO WORK WITH ME.
Write in standard form.
fifty-nine thousand, eight hundred

YOUR TURN #1:   **YOUR TURN #2:**

YOUR TURN #3:   **YOUR TURN #4:**

Objective C: Write a Whole Number in Expanded Form

Watch the objective video.

VIDEO WORK WITH ME.
Write in expanded form:
80,774

The _____ of a number shows each digit of the number with its place value.

YOUR TURN #1:

Objective D: Read Tables

Watch the objective video.

VIDEO WORK WITH ME.

Which breed has a greater average dog maximum weight, the Bulldog or the German Shepherd?

Which breed is the most popular dog? What is the maximum weight of that dog written in words?

What is the maximum weight of an average-size Boxer?

YOUR TURN #1:   **YOUR TURN #2:**

Section 1.3 Adding and Subtracting Whole Numbers and Perimeter

Objectives
>   A Add Whole Numbers
>   B Subtract Whole Numbers
>   C Find the Perimeter of a Polygon
>   D Solve Problems by Adding or Subtracting Whole Numbers

Directions: Complete your Interactive Organizer by filling in the blanks and solving exercises as you complete each screen of the Interactive Assignment.

- For **WORK WITH ME** exercises, follow along and write each step needed and shown to solve, including the final answer.
- For **YOUR TURN** exercises, write the exercise generated for you in MyLab Math, then "show your work" by writing each step needed to solve, including the final answer.

Objective A: Add Whole Numbers

Each of the numbers 2 and 4 is called an _____ , and the process of finding the _____ (or total) is called _____ .

The Process of Addition

$$2 \quad + \quad 4 \quad = \quad 6$$

$$\uparrow \qquad\qquad \uparrow \qquad\qquad \uparrow$$

_____   addend   _____

To add whole numbers, we add the digits in the _____ place, then the _____ place, then the _____ place, and so on.

$$\begin{array}{r} 2\ 2\ 3\ 6 \\ +\ \ \ 1\ 6\ 0 \\ \hline \end{array}$$

Let's review some properties of addition.

Addition Property of Zero
>   The sum of 0 and any number is that _____ .
>   For example,

Commutative Property of Addition
>   Changing the _____ of two addends does not change their sum.
>   For example,

Associative Property of Addition
Changing the _____ of addends does not change their sum.
For example,

WORK WITH ME #1.

Add.

a.   5 2 6 7
   +   1 3 2

b. 24 + 9006 + 489 + 2407

YOUR TURN #1:

YOUR TURN #2:

Objective B: Subtract Whole Numbers

Below 8 is called the _____ , and 3 is the _____ .The_____ between these two numbers, 8 and 3, is 5.

The Process of Subtraction

      8   –   3   =   5
      ↑   ↑   ↑

   _____   subtrahend   _____

Addition and subtraction are very closely related.
   8 – 3 = 5 because

   Subtraction can be checked by addition, and we say that addition and subtraction are _____ operations.

Subtraction Properties of 0
The difference of any number and that _____ number is 0.
For example,

The difference of any number and 0 is that _____ number.
For example,

To subtract whole numbers, we subtract the digits in the _____ place, then the _____ place, then the _____ place, and so on.

Show Me   **Now Check**

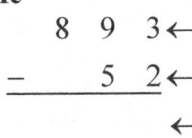

WORK WITH ME #1.

Subtract. Check by adding

  a.   7 4 9   b.   6 2
      − 1 4 9   − 3 7

  c. 51,111 − 19,898

YOUR TURN #1:   **YOUR TURN #2:**

Objective C: Find the Perimeter of a Polygon

Watch the objective video.

  The _____ of a polygon is the distance around the polygon.

***VIDEO* WORK WITH ME.**

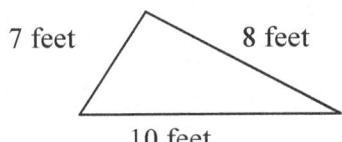

7 feet   8 feet

10 feet

YOUR TURN #1:

Objective D: Solve Problems by Adding or Subtracting Whole Numbers

Watch the objective video.

VIDEO WORK WITH ME.

A stereo that regularly sells for $547 is discounted by $99 in a sale. What is the sale price?

regular price $\rightarrow$
$-$ discount
       sale price

YOUR TURN #1:   **YOUR TURN #2:**

Section 1.4 Rounding and Estimation

Objectives
>   A Round Whole Numbers
>   B Use Rounding to Estimate Sums and Differences
>   C Solve Problems by Estimating

Directions: Complete your Interactive Organizer by filling in the blanks and solving exercises as you complete each screen of the Interactive Assignment.
- For **WORK WITH ME** exercises, follow along and write each step needed and shown to solve, including the final answer.
- For **YOUR TURN** exercises, write the exercise generated for you in MyLab Math, then "show your work" by writing each step needed to solve, including the final answer.

Objective A: Round Whole Numbers

Watch the objective video.

_____ a whole number means approximating it.

Rounding Whole Numbers to a Given Place Value

Step 1:

Step 2:

Step 3:

VIDEO WORK WITH ME.

Round 635 to the nearest ten.   Round 36,499 to the nearest thousand.

WORK WITH ME #1.

YOUR TURN #1:   **YOUR TURN #2:**

9

Objective B: Use Rounding to Estimate Sums and Differences

By rounding addends, minuends, and subtrahends, we can _____ sums and differences.

This is appropriate when the _____ number is not needed or to help us determine if we made a mistake in calculating an exact amount.

WORK WITH ME #1.

WORK WITH ME #2.

a. Estimate the sum by rounding
each number to the nearest ten

$$
\begin{array}{r}
39 \\
45 \\
22 \\
+17 \\
\hline
\end{array}
$$

b. Estimate the sum by rounding
each number to the nearest hundred

$$
\begin{array}{r}
1774 \\
-1492 \\
\hline
\end{array}
$$

YOUR TURN #1:

YOUR TURN #2:

Objective C: Solve Problems by Estimating

Watch the objective video.

VIDEO WORK WITH ME.
The peak of Denali (Mt. McKinley), in Alaska, is 20,320 feet above sea level. The top of Mt. Rainier, in Washington, is 14,410 feet above sea level. Round each height to the nearest thousand to estimate the difference in elevation of these two peaks.

YOUR TURN #1:

YOUR TURN #2:

Section 1.5 Multiplying Whole Numbers and Area

A Use Properties of Multiplication
B Multiply Whole Numbers
C Find the Area of a Rectangle
D Solve Problems by Multiplying Whole Numbers

Directions: Complete your Interactive Organizer by filling in the blanks and solving exercises as you complete each screen of the Interactive Assignment.

- For **WORK WITH ME** exercises, follow along and write each step needed and shown to solve, including the final answer.
- For **YOUR TURN** exercises, write the exercise generated for you in MyLab Math, then "show your work" by writing each step needed to solve, including the final answer.

Objective A: Use Properties of Multiplication

_____ is repeated addition of the same number.

WORK WITH ME #1.

Let's review some properties of multiplication.

Multiplication Property of 0
    The product of 0 and any number is _____ .
    For example,

Multiplication Property of 1
    The product of 1 and any number is that _____ number.
    For example,

Let's review more properties of multiplication.

Commutative Property of Multiplication
    Changing the _____ of two factors does not change their product.
    For example,

Associative Property of Multiplication
    Changing the _____ of factors does not change their product.
    For example,

11

Now let's review the distributive property.

> **Distributive Property**
>   Multiplication distributes over _____ .
>   For example,

WORK WITH ME #2.

a. Multiply.
   $1 \cdot 24$

b. Multiply.
   $0 \cdot 19$

c. Use the distributive property to rewrite the expression.
   $20(14 + 6)$

YOUR TURN #1:

YOUR TURN #2:

Objective B: Multiply Whole Numbers

Watch the objective video.

VIDEO WORK WITH ME.

$$\begin{array}{r} 277 \\ \times \ \ 6 \\ \hline \end{array}$$

$$\begin{array}{r} 8\,6\,4\,9 \\ \times \ 2\,7\,4 \\ \hline \end{array}$$

YOUR TURN #1:

YOUR TURN #2:

Objective C: Find the Area of a Rectangle

Watch the objective video.

VIDEO WORK WITH ME.

_____ measures the surface of a region.   Area =

9 meters

```
┌─────────────────┐
│                 │  7 meters
│                 │
└─────────────────┘
```

_____ is measured in square units.

YOUR TURN #1:

Objective D: Solve Problems by Multiplying Whole Numbers

Watch the objective video.

VIDEO WORK WITH ME.

One tablespoon of olive oil contains 125 calories. How many calories are in 3 tablespoons of olive oil?

YOUR TURN #1:

Section 1.6 Dividing Whole

Objectives

　　　A　Dividing Whole Numbers
　　　B　Perform Long Division
　　　C　Solve Problems That Require Dividing by Whole Numbers
　　　D　Find the Average of a List of Numbers

Directions: Complete your Interactive Organizer by filling in the blanks and solving exercises as you complete each screen of the Interactive Assignment.
- For **WORK WITH ME** exercises, follow along and write each step needed and shown to solve, including the final answer.
- For **YOUR TURN** exercises, write the exercise generated for you in MyLab Math, then "show your work" by writing each step needed to solve, including the final answer.

Objective A:　Dividing Whole Numbers

Just as subtraction is the reverse of addition, division is the reverse of _____ . Thus, division can be checked by _____ .

Here are some ways to write division.

　One Way　　　　　　　　　　　　　　　　Another Way

　Yet Another Way

Let's review some properties of division.

Division Properties of 1

　　　The quotient of any number (except 0) and that same number is _____ .

　　　The quotient of any number and 1 is the _____ number.

Division Properties of 0

The quotient of 0 and any number (except 0) is _____ .

The quotient of any number and 0 is not a _____ .

WORK WITH ME #1.

Find each quotient.

a. $36 \div 3$

b. $31 \div 1$

c. $\dfrac{18}{18}$

d. $0 \div 14$

e. $26 \div 0$

YOUR TURN #1:   **YOUR TURN #2:**

Objective B: Perform Long Division

Watch the objective video.

VIDEO WORK WITH ME.

$55\overline{)715}$ $20,619 \div 102$

YOUR TURN #1:   **YOUR TURN #2:**

Objective C: Solve Problems That Require Dividing by Whole Numbers

Watch the objective video.

***VIDEO* WORK WITH ME.**
An 18-hole golf course is 5580 yards long. If the distance to each hole is the same, find the distance between holes.

YOUR TURN #1:

Objective D: Find the Average of a List of Numbers

Watch the objective video.

Average of a List of Numbers

$$\text{average} = \frac{\text{sum of numbers}}{\textit{number} \text{ of numbers}}$$

***VIDEO* WORK WITH ME.**
Find the average: 86, 79, 81, 69, 80

WORK WITH ME #1.

YOUR TURN #1:

Section 1.7 Exponents and Order of Operations

Objectives
A　Write Repeated Factors Using Exponential Notation B　Evaluate Expressions Containing Exponents C　Use the Order of Operations D　Find the Area of a Square

Directions: Complete your Interactive Organizer by filling in the blanks and solving exercises as you complete each screen of the Interactive Assignment.
- For **WORK WITH ME** exercises, follow along and write each step needed and shown to solve, including the final answer.
- For **YOUR TURN** exercises, write the exercise generated for you in MyLab Math, then "show your work" by writing each step needed to solve, including the final answer.

Objective A: Write Repeated Factors Using Exponential Notation

Watch the objective video.

A shorthand notation for repeated multiplication of the same factor is called _____ notation.

VIDEO WORK WITH ME.
$12 \cdot 12 \cdot 12$　　　　　　　　　　　　　$6 \cdot 6 \cdot 5 \cdot 5 \cdot 5$

YOUR TURN #1:

Objective B: Evaluate Expressions Containing Exponents

WORK WITH ME #1.
Evaluate exponential expressions.

a. 6^4　　　　　　　　b. 5^1　　　　　　　　c. $3^2 \cdot 4^3$

Helpful Hint
An exponent applies only to its _____ . Don't forget that 2^4 is *not* _____ . The expression 2^4 means repeated multiplication of the same _____ .

YOUR TURN #1:   **YOUR TURN #2:**

Objective C: Use the Order of Operations

Watch the objective video.

VIDEO WORK WITH ME.
Evaluate:

$15 + 3 \cdot 2$

Order of Operations

Step 1:

Step 2:

Step 3:

Step 4:

VIDEO WORK WITH ME.
Evaluate:

$14 \div 7 \cdot 2 + 3$ $\dfrac{7(9-6)+3}{3^2 - 3}$

$(7 \cdot 5) + [9 \div (3 \div 3)]$

YOUR TURN #1:   **YOUR TURN #2:**

YOUR TURN #3:

Objective D: Find the Area of a Square

Watch the objective video.

Area of rectangle =   Area of square =

***VIDEO* WORK WITH ME.**

7 meters

YOUR TURN #1:

Section 1.8 Introduction to Variables, Algebraic Expressions, and Equations

Objectives
A Evaluate Algebraic Expressions Given Replacement Values
B Identify Solutions of Equations
C Translate Phrases into Variable Expressions

Directions: Complete your Interactive Organizer by filling in the blanks and solving exercises as you complete each screen of the Interactive Assignment.

- For **WORK WITH ME** exercises, follow along and write each step needed and shown to solve, including the final answer.
- For **YOUR TURN** exercises, write the exercise generated for you in MyLab Math, then "show your work" by writing each step needed to solve, including the final answer.

Objective A: Evaluate Algebraic Expressions Given Replacement Values

A letter used to represent a number is called a _____ .

Variables and numbers combined by operations are called _____ _____, or simply _____ .

$2x$ means _____

xy means _____

VIDEO WORK WITH ME.

If $x = 2$, $y = 5$, and $z = 3$, evaluate:

$3 + 2z =$ $y^3 - 4x =$

$2y(4z - x) =$

YOUR TURN #1:   **YOUR TURN #2:**

YOUR TURN #3:

Objective B: Identify Solutions of Equations

An equation is of the form _____ = _____ .

Helpful Hint
An equation contains "=," while an expression does not.

How do we label an equation?

A _____ of an equation containing a variable is a value for the variable that makes an equation
a _____ statement.

WORK WITH ME #1.

WORK WITH ME #2.   **YOUR TURN #1:**
Is 8 a solution of $7f = 64 - f$?

Objective C: Translate Phrases into Variable Expressions

Watch the objective video.

Addition (+)	Subtraction (−)	Multiplication (·)	Division (÷)

VIDEO WORK WITH ME.

Twenty decreased by a number   The product of five and a number

---

Helpful Hint

Remember that order is important when subtracting. Study the order of numbers and variables below.

Phrase	Translation
A number *decreased* by 5	
A number *subtracted from* 5	

YOUR TURN #1:   **YOUR TURN #2:**

YOUR TURN #3:

Chapter 1 Review and Practice

> Study Skills
> Chapter Vocabulary
> Getting Ready for the Test
> Review Exercises
> Practice Chapter Test

Study Skills

Directions: **Watch the Study Skills video.**

Chapter Vocabulary

WORK WITH ME.

Fill in each blank with one of the words or phrases listed below:

| difference | factor | perimeter | dividend | minuend | whole numbers |

| equation | divisor | variable | sum | addend | exponent | expression |

| quotient | subtrahend | product | digits | area |

1. The _____ are 0, 1, 2, 3, …

2. The _____ of a polygon is its distance around or the sum of the lengths of its sides.

3. A(n) _____ is shorthand notation for repeated multiplication of the same factor.

4. To find the _____ of a rectangle, multiply length times width.

5. The _____ used to write numbers are 0, 1, 2, 3, 4, 5, 6, 7, 8, and 9.

6. A letter used to represent a number is called a(n) _____ .

7. A(n) _____ can be written in the form "expression = expression."

Use the facts below for Exercises 8 through 14.

$$2 \cdot 3 = 6 \qquad 4 + 17 = 21 \qquad 20 - 9 = 11 \qquad 5\overline{)35}\;^{7}$$

8. The 21 above is called the _____ .

9. The 5 above is called the _____ .

10. The 7 above is called the _____ .

11. The 3 above is called a(n) _____ .

12. The 6 above is called the _____ .

13. The 11 above is called the _____ .

14. The 4 above is called a(n) _____ .

Getting Ready for the Test.

- These exercises will help you avoid common errors while taking your chapter test.

General Directions: Read the exercise Write any notes or steps in this Interactive Organizer, along with your answer to the exercise. In the MyLab Math Interactive Assignment, click the **SHOW ANSWERS** button to check your answers. Correct any errors, or press the **PLAY** button for a video solution.

Multiple Choice. *For Exercises 1 through 4, name the place value for the given digit. The number is 28,690,357,004 and choices are:*

  A. tens   **B.** millions   **C.** ten-millions   **D.** ten-thousands   **E.** billions   **F.** hundred-millions

1. the digit 5

2. the digit 8

3. the digit 6

4. the digit 0 to the far left

Multiple Choice. *For Exercises 5 through 8, identify the first operation to be performed to simplify the expression.*

  A. add   **B.** subtract   **C.** multiply   **D.** divide

5. $6 - 3 \cdot 2$

6. $(6 - 3) \cdot 2$

7. $6 \div 3 \cdot 2$

8. $6 + 3 - 2$

9. The expression $5 \cdot 2^3$ simplifies to
       A. 1000   B. 30   C. 40   D. 13

Multiple Choice. For Exercises 10 through 13, let $a = 35$ and $b = 5$. Choose the expression that gives each answer.

  A. $a - b$ **B.** $a \div b$ **C.** $a + b$ **D.** ab

10. answer: 7

11. answer: 175

12. answer: 30

13. answer: 40

Review Exercises

In the **MyLab Math, Interactive Assignment, Review Exercises** section, there are algorithmically generated "Your Turn" exercises so that you can check your knowledge of some core concepts in this chapter. Insert a few sheets of paper in your Interactive Organizer to "record and show your work" along with the final answer.

Practice Chapter Test

- These exercises will help you practice for your chapter test.

General Directions: For each exercise, "show your work" by writing each step in the solution process within your Interactive Organizer, including your final answer. In the MyLab Math Interactive Assignment, click the Show Answer button to check your answer. Correct any errors, or press the **PLAY** button for a video solution.

For Exercises 1 through 15, simplify.

1. Write 82,426 in words.

2. Write "four hundred two thousand, five hundred fifty" in standard form.

3. $59 + 82$

4. $600 - 487$

5. $\begin{array}{r} 496 \\ \times\ 30 \\ \hline \end{array}$

6. $52,896 \div 69$

7. $2^3 \cdot 5^2$

8. $98 \div 1$

9. $0 \div 49$

10. $62 \div 0$

11. $(2^4 - 5) \cdot 3$

12. $16 + 9 \div 3 \cdot 4 - 7$

13. $6^1 \cdot 2^3$

14. $2[(6-4)^2 + (22-19)^2] + 10$

15. $5698 \cdot 1000$

16. Find the average of 62, 79, 84, 90, and 95.

17. Round 52,369 to the nearest thousand.

For Exercises 18 through 18, estimate each sum or difference by rounding each number to the nearest hundred.

18. $6289 + 5403 + 1957$

19. $4267 - 2738$

For Exercises 20 through 27, solve.

20. Subtract 15 from 107.

21. Find the sum of 15 and 107.

22. Find the product of 15 and 107.

23. Find the quotient of 107 and 15.

24. Twenty-nine cans of Sherwin-Williams paint cost $493. How much was each can?

25. Jo McElory is looking at two new refrigerators for her apartment. One costs $599 and the other costs $725. How much more expensive is the higher-priced one?

26. One tablespoon of white granulated sugar contains 45 calories. How many calories are in 8 tablespoons of white granulated sugar? (*Source* : *Home and Garden Bulletin No.* 72, U.S. Department of Agriculture)

27. A small business owner recently ordered 16 digital cameras that cost $430 each and 5 printers that cost $205 each. Find the total cost for these items.

For Exercises 28 through 29, find the perimeter and the area of each figure.

28.

| Square | 5 centimeters |

29.

20 yards

| Rectangle | 10 yards |

30. Evaluate $5(x^3 - 2)$ for $x = 2$.

31. Evaluate $\dfrac{3x-5}{2y}$ for $x = 7$ and $y = 8$.

32. Translate the following phrases into mathematical expressions. Use x to represent "a number."
    a. The quotient of a number and 17
    b. Twice a number, decreased by 20

33. Is 6 a solution of the equation $5n - 11 = 19$?

34. Determine which number in the set is a solution to the given equation.
    $n + 20 = 4n - 10$; {0, 10, 20}

27

Section 2.1 Introduction to Integers

Objectives
A Represent Real-Life Situations with Integers
B Graph Integers on a Number Line
C Compare Integers
D Find the Absolute Value of a Number
E Find the Opposite of a Number
F Read Bar Graphs Containing Integers

Directions: Complete your Interactive Organizer by filling in the blanks and solving exercises as you complete each screen of the Interactive Assignment.

- For **WORK WITH ME** exercises, follow along and write each step needed and shown to solve, including the final answer.
- For **YOUR TURN** exercises, write the exercise generated for you in MyLab Math, then "show your work" by writing each step needed to solve, including the final answer.

Objective A: Represent Real-Life Situations with Integers

Watch the objective video.

_____ Numbers: 0, 1, 2, …

When drawing a number line we have: _____ numbers, _____ , Positive numbers

VIDEO WORK WITH ME.

A worker in a _____ mine in Nevada works _____ feet underground.

Represent the quantity as an integer. _____

YOUR TURN #1:

Objective B: Graph Integers on a Number Line

Watch the objective video.

VIDEO WORK WITH ME.

Graph: –1, 1, –2, 2, –4 on a number line.

YOUR TURN #1:

Objective C: Compare Integers

The inequality symbol > means _____ .

The inequality symbol < means _____ .

Both –5 and –7 are graphed on the number line below.

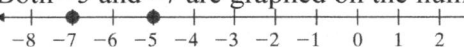

–7 ____ –5 (since –7 is to the _____ of –5 on the number line)

–5 _____ –7 (since –5 is to the _____ of –7 on the number line)

Helpful Hint
If you think of < and > as _____ , notice that in a true statement, the arrow always points to the _____ number.
5 > _____   –3 < _____

YOUR TURN #1:

Objective D: Find the Absolute Value of a Number

The _____ of a number is the number's distance from 0 on the number line.

The symbol for absolute value is _____ .

$|3| = 3$ because 3 is ____ units from ____ .

Draw the graph.

$|{-3}| = 3$ because –3 is ____ units from ____ .

Draw the graph.

Helpful Hint
Since the absolute value of a number is that number's _____ from 0, the absolute value of a number is always 0 or _____ . It is never _____ .
$\mid 0 \mid =$ _____ $|{-6}| =$ _____

YOUR TURN #1:   **YOUR TURN #2:**

Objective E: Find the Opposite of a Number

Two numbers that are the same distance from 0 on the number line but are on opposite sides of 0 called _____ .

   4 and _____ are opposites.

   Draw the graph.

WORK WITH ME #1.

Simplify.
   a. Find the opposite of –4.
   b. $-|-3|$
   c. $-|20|$
   d. $-(-33)$

YOUR TURN #1:   **YOUR TURN #2:**

Objective F: Read Bar Graphs Containing Integers

Watch the objective video.

VIDEO WORK WITH ME.

Which lake shown has the lowest elevation?

What is the elevation?

VIDEO WORK WITH ME.

Which lake shown has the highest elevation?

What is the elevation?

   The absolute value of a number is that number's distance from _____ on the number line.

   Two numbers are _____ if they lie on opposite sides of 0 on the number line and the same distance from 0.

YOUR TURN #1:

Section 2.2 Adding Integers

Objectives

    A Add Integers

    B Evaluate an Algebraic Expression by Adding

    C Solve Problems by Adding Integers

Directions: Complete your Interactive Organizer by filling in the blanks and solving exercises as you complete each screen of the Interactive Assignment.

- For **WORK WITH ME** exercises, follow along and write each step needed and shown to solve, including the final answer.
- For **YOUR TURN** exercises, write the exercise generated for you in MyLab Math, then "show your work" by writing each step needed to solve, including the final answer.

Objective A: Add Integers

WORK WITH ME #1.

Add using a number line. $-1 + (-6)$

Adding Two Numbers with the Same Sign

    Step 1. _____ their absolute values.

    Step 2. Use their _____ sign as the sign of the sum.

WORK WITH ME #2.

Add. $-2 + (-7)$

YOUR TURN #1:

WORK WITH ME #3.

Add. $-4 + 7$

Adding Two Numbers with Different Signs

Step 1. Find the larger absolute value _____ the smaller absolute value.

Step 2. Use the sign of the number with the _____ absolute sign as the sign of the sum.

WORK WITH ME #4.

Add. $-6 + 3$

YOUR TURN #2:   **YOUR TURN #3:**

WORK WITH ME #5.

Add. $-4 + 2 + (-5)$

YOUR TURN #4:

Objective B: Evaluate an Algebraic Expression by Adding

WORK WITH ME #1.

Evaluate $3x + y$ for the given replacement values. $x = 2$ and $y = -3$

YOUR TURN #1:

Objective C: Solve Problems by Adding Integers

WORK WITH ME #1.

Solve.
Suppose a deep-sea diver dives from the surface to 215 feet below the surface. He then dives down 16 more feet. Use positive and negative numbers to represent this situation. Then find the diver's present depth.

YOUR TURN #1:

Section 2.3 Subtracting Integers

Objectives
    A Subtract Integers
    B Add and Subtract Integers
    C Evaluate an Algebraic Expression by Subtracting
    D Solve Problems by Subtracting Integers

Directions: Complete your Interactive Organizer by filling in the blanks and solving exercises as you complete each screen of the Interactive Assignment.
- For **WORK WITH ME** exercises, follow along and write each step needed and shown to solve, including the final answer.
- For **YOUR TURN** exercises, write the exercise generated for you in MyLab Math, then "show your work" by writing each step needed to solve, including the final answer.

Objective A: Subtract Integers

To subtract two numbers, we add the first number to the _____ of the second number.

Subtracting Two Numbers
    If a and b are numbers, then $a - b =$

WORK WITH ME #1.
Subtract.
    a. $3 - 8$

    b. $-7 - (-3)$

Helpful Hint
    To visualize subtraction, try the following:
        The difference between 5°F and –2°F can be found by subtracting. That is,
            $5 - (-2) =$

YOUR TURN #1:   **YOUR TURN #2:**

WORK WITH ME #2.   **YOUR TURN #3:**
Translate the phrase; then simplify.
    Subtract 17 from –25.

Objective B: Add and Subtract Integers

WORK WITH ME #1.

$7 - 8 - (-5) - 1$

YOUR TURN #1:

Objective C: Evaluate an Algebraic Expression by Subtracting

WORK WITH ME #1.

Evaluate $2x - y$ for the given replacement values.
   $x = 4$ and $y = -4$

YOUR TURN #1:

Objective D: Solve Problems by Subtracting Integers

Watch the objective video.

***VIDEO* WORK WITH ME.**

The coldest temperature ever recorded on Earth was $-129°F$ in Antarctica. The warmest temperature ever recorded was $134°F$ in Death Valley, California. How many degrees warmer is $134°F$ than $-129°F$?

YOUR TURN #1:

Section 2.4 Multiplying and Dividing Integers

Objectives

A Multiply Integers
B Divide Integers
C Evaluate an Algebraic Expression by Multiplying or Dividing
D Solve Problems by Multiplying or Dividing Integers

Directions: Complete your Interactive Organizer by filling in the blanks and solving exercises as you complete each screen of the Interactive Assignment.

- For **WORK WITH ME** exercises, follow along and write each step needed and shown to solve, including the final answer.
- For **YOUR TURN** exercises, write the exercise generated for you in MyLab Math, then "show your work" by writing each step needed to solve, including the final answer.

Objective A: Multiply Integers

Multiplying and dividing integers is similar to multiplying and dividing whole numbers. One difference is that we need to determine whether the result is a positive number or a negative number.

Multiplying Numbers

The product of two numbers having the _____ sign is a positive number.
    $(+)(+) =$
    $(-)(-) =$

The product of two numbers having _____ signs is a negative number.
    $(-)(+) =$
    $(+)(-) =$

WORK WITH ME #1.

Multiply.

a. $-4(9)$ b. -3^2 c. $-3(-4)(-2)$

Helpful Hint

Have you noticed a pattern when _____ signed numbers?

Let $(-)$ represent a negative number and $(+)$ represent a positive number.

Product of Even Integers
  The product of an even number of
  negative numbers is a _____ result.
    $(-)(-) =$
    $(-)(-)(-)(-) =$

Product of Odd Integers
  The product of an odd number of
  negative numbers is a _____ result.
    $(-)(-)(-) =$
    $(-)(-)(-)(-)(-) =$

YOUR TURN #1:   **YOUR TURN #2:**

YOUR TURN #3:   **YOUR TURN #4:**

Objective B: Divide Integers

The sign rules for division can be discovered and checked by writing a related multiplication problem.

$$\frac{6}{2} = 3 \qquad \text{Why?}$$

Helpful Hint Just as for whole numbers, division can be checked by _____ .

$$\frac{-6}{2} = -3 \qquad \text{Why?}$$

$$\frac{6}{-2} = -3 \qquad \text{Why?}$$

$$\frac{-6}{-2} = 3 \qquad \text{Why?}$$

Dividing Numbers
The quotient of two numbers having the same sign is a _____ number.

$$\frac{(+)}{(+)} = \qquad\qquad \frac{(-)}{(-)} =$$

The quotient of two numbers having different signs is a _____ number.

$$\frac{(+)}{(-)} = \qquad\qquad \frac{(-)}{(+)} =$$

WORK WITH ME #1.

a. $\dfrac{-30}{6}$ b. $\dfrac{-12}{-4}$ c. $\dfrac{0}{-15}$ d. $\dfrac{-10}{0}$

YOUR TURN #1:   **YOUR TURN #2:**

YOUR TURN #3:   **YOUR TURN #4:**

Objective C: Evaluate an Algebraic Expression by Multiplying or Dividing

Watch the objective video.

VIDEO WORK WITH ME.

a. Evaluate *ab* for $a = 9$ and $b = -2$.   b. Evaluate $\dfrac{x}{y}$ for $x = 5$ and $y = -5$.

YOUR TURN #1:

Objective D: Solve Problems by Multiplying or Dividing Integers

Watch the objective video.

VIDEO WORK WITH ME.

A football team lost four yards on each of three consecutive plays. Represent the total loss as a product of signed numbers and find the total loss.

YOUR TURN #1:   **YOUR TURN #2:**

Section 2.5 Order of Operations

Objectives

    A Simplify Expressions by Using the Order of Operations
    B Evaluate an Algebraic Expression
    C Find the Average of a List of Numbers

Directions: Complete your Interactive Organizer by filling in the blanks and solving exercises as you complete each screen of the Interactive Assignment.

- For **WORK WITH ME** exercises, follow along and write each step needed and shown to solve, including the final answer.
- For **YOUR TURN** exercises, write the exercise generated for you in MyLab Math, then "show your work" by writing each step needed to solve, including the final answer.

Objective A: Simplify Expressions by Using the Order of Operations

See if you can remember the order of operations.

Order of Operations
Step 1.
Step 2.
Step 3.
Step 4.

WORK WITH ME #1.
What operation is performed first when evaluating $7 + 3(9 - 6 \cdot 2)$?

WORK WITH ME #2.
Simplify.

    a. $4 - (-3)^4$ b. $-3[5 + 2(-4 + 9)] + 15$

WORK WITH ME #3.
    a. $(-8)^2$ b. -8^2

YOUR TURN #1:   **YOUR TURN #2:**

YOUR TURN #3:

Objective B: Evaluate an Algebraic Expression

WORK WITH ME #1.
Evaluate the expression for $x = -2$ and $y = 4$.
$$x^2 - y$$

WORK WITH ME #2.
Evaluate the expression for $x = -3$ and $z = -4$.
$$2x^3 - z$$

YOUR TURN #1:   **YOUR TURN #2:**

Objective C: Find the Average of a List of Numbers

Watch the objective video.

Recall

Average of a List of Numbers
$$\text{average} = \frac{\text{sum of numbers}}{\textit{number} \text{ of numbers}}$$

***VIDEO* WORK WITH ME.**

$-10, 8, -4, 2, 7, -5, -12$

YOUR TURN #1:

Section 2.6 Solving Equations: The Addition and Multiplication Properties

Objectives
   A Identify Solutions of Equations
   B Use the Addition Property of Equality to Solve Equations
   C Use the Multiplication Property of Equality to Solve Equations

Directions: Complete your Interactive Organizer by filling in the blanks and solving exercises as you complete each screen of the Interactive Assignment.
   - For **WORK WITH ME** exercises, follow along and write each step needed and shown to solve, including the final answer.
   - For **YOUR TURN** exercises, write the exercise generated for you in MyLab Math, then "show your work" by writing each step needed to solve, including the final answer.

Objective A: Identify Solutions of Equations

Watch the objective video.

Equation: "expression" = "expression"   An _____ has "=", an expression does not.

   A number is a _____ of an equation if a true statement results when the variable is replaced by the number.

VIDEO WORK WITH ME.
Determine whether −5 is a solution of $x + 12 = 17$.

_____ statement: the number *is a solution.*   _____ statement: the number *is not a solution.*

YOUR TURN #1:

Objective B: Use the Addition Property of Equality to Solve Equations

To solve an equation, we write simpler equations, all _____ to the original equation, until the final equation has the form
        $x =$ number   or   number $= x$

The _____ **property of equality** helps us write equivalent equations.

Let a, b, and c represent numbers. Then

$a = b$	Also, $a = b$
and $a + c = $ _____	and $a - c = $ _____
are equivalent equations.	are equivalent equations.

WORK WITH ME #1.
Use the Animated Scale to solve
$$x - 2 = 1.$$

WORK WITH ME #2.
Solve. Check each solution.

a. $a + 5 = 23$

b. $7 = y - 2$

c. $-7 + 10 - 20 = x$

YOUR TURN #1:   **YOUR TURN #2:**

YOUR TURN #3:

Objective C: Use the Multiplication Property of Equality to Solve Equations

A second property of equality is called the _____ **property of equality.**

Let a, b, and c represent numbers and let $c \neq 0$. Then

$a = b$	Also, $a = b$
and $a \cdot c =$ _____	and $\dfrac{a}{c} =$ _____
are equivalent equations.	are equivalent equations.

WORK WITH ME #1.
Use the Animated Scale to solve
$$2x = 6.$$

WORK WITH ME #2.
Solve. Check each solution.

a. $5x = 20$

b. $-3z = 12$

c. $\dfrac{n}{4} = -20$

YOUR TURN #1:

YOUR TURN #2:

Chapter 2 Review and Practice

> Study Skills
> Chapter Vocabulary
> Getting Ready for the Test
> Review Exercises
> Practice Chapter Test

Study Skills

Directions: **Watch the Study Skills video.**

Chapter Vocabulary

WORK WITH ME.

Directions: Fill in each blank with one of the words or phrases listed below:

| inequality symbols | addition | solution | is less than | integers | expression |

| average | negative | absolute value | equation | positive | opposites |

| is greater than | multiplication |

1. Two numbers that are the same distance from 0 on the number line but are on opposite sides of 0 are called _____ .

2. The _____ of a number is that number's distance from 0 on a number line.

3. The _____ are … , –3, –2, –1, 0, 1, 2, 3, ….

4. The _____ numbers are less than zero.

5. The _____ numbers are greater than zero.

6. The symbols "<" and ">" are called _____ .

7. A(n) _____ of an equation is a number that when substituted for a variable makes the equation a true statement.

8. The _____ of a list of numbers is $\dfrac{\text{sum of numbers}}{\textit{number} \text{ of numbers}}$.

9. A combination of operations on variables and numbers is called a(n) _____ .

10. A statement of the form "expression = expression" is called a(n) _____ .

11. The sign "<" means _____ and ">" means _____ .

12. By the _____ property of equality, the same number may be added to or subtracted from both sides of an equation without changing the solution of the equation.

13. By the _____ property of equality, both sides of an equation may be multiplied or divided by the same nonzero number without changing the solution of the equation.

Getting Ready for the Test

- These exercises will help you avoid common errors while taking your chapter test.

General Directions: Read the exercise Write any notes or steps in this Interactive Organizer, along with your answer to the exercise. In the MyLab Math Interactive Assignment, click the **SHOW ANSWERS** button to check your answers. Correct any errors, or press the **PLAY** button for a video solution.

Multiple Choice. *For Exercises 1 through 4, identify each answer as*

  A. 2   **B.** −2   **C.** 0

1. −2; choose the opposite

2. −2; choose the absolute value

3. 2; choose the absolute value

4. 2; choose the opposite

Multiple Choice. *For Exercises 5 through 10, identify each answer as*

  A. addition   **B.** subtraction   **C.** multiplication   **D.** division

5. For xy, the operation is _____ .

6. For $-12(+3)$, the operation is _____ .

7. For $-12 + 3$, the operation is _____ .

8. For $\dfrac{-12}{+3}$, the operation is _____ .

9. For $4 + 6 \cdot 2$, which operation is performed first?

10. For $4 + 6 \div 2$, which operation is performed first?

Multiple Choice. *For Exercises 11 through 14, choose the solution of each equation as*

   A. 2   **B.** −2   **C.** 18   **D.** −18

11. $-3x = 6$

12. $x - 2 = -4$

13. $\dfrac{x}{-3} = 6$

14. $x + 2 = 4$

Review Exercises

In the **MyLab Math, Interactive Assignment, Review Exercises** section, there are algorithmically generated "Your Turn" exercises so that you can check your knowledge of some core concepts in this chapter. Insert a few sheets of paper in your Interactive Organizer to "record and show your work" along with the final answer.

Practice Chapter Test

- These exercises will help you practice for your chapter test.

General Directions: For each exercise, "show your work" by writing each step in the solution process within your Interactive Organizer, including your final answer. In the MyLab Math Interactive Assignment, click the Show Answer button to check your answer. Correct any errors, or press the **PLAY** button for a video solution.

For Exercises 1 through 22, simplify each expression.

1. $-5 + 8$ 2. $18 - 24$ 3. $5 \cdot (-20)$

4. $-16 \div (-4)$ 5. $-18 + (-12)$ 6. $-7 - (-19)$

7. $-5 \cdot (-13)$ 8. $\dfrac{-25}{-5}$ 9. $|-25| + (-13)$

10. $14 - |-20|$

11. $|5| \cdot |-10|$

12. $\dfrac{|-10|}{-|-5|}$

13. $-8 + 9 \div (-3)$

14. $-7 + (-32) - 12 + 5$

15. $(-5)^3 - 24 \div (-3)$

16. $(5-9)^2 \cdot (8-2)^3$

17. $-(-7)^2 \div 7 \cdot (-4)$

18. $3 - (8-2)^3$

19. $\dfrac{4}{2} - \dfrac{8^2}{16}$

20. $\dfrac{-3(-2) + 12)}{-1(-4 - 5)}$

21. $\dfrac{|25 - 30|^2}{2(-6) + 7}$

22. $5(-8) - [6 - (2-4)] + (12 - 16)^2$

For Exercises 23 through 25, evaluate each expression for x = 0, y = –3, and z = 2.

23. $7x + 3y - 4z$

24. $10 - y^2$

25. $\dfrac{3z}{2y}$

26. Mary Dunstan, a diver, starts at sea level and then makes 4 successive descents of 22 feet. After the descents, what is her elevation?

27. Aaron Hawn has $129 in his checking account. He writes a check for $79, withdraws $40 from an ATM, and then deposits $35. Represent the new balance in his account by an integer.

28. Mt. Washington in New Hampshire has an elevation of 6288 feet above sea level. The Romanche Gap in the Atlantic Ocean has an elevation of 25,354 feet below sea level. Represent the difference in elevation between these two points by an integer. (*Source*: National Geographic Society and Defense Mapping Agency)

29. Lake Baykal in Siberian Russia is the deepest lake in the world, with a maximum depth of 5315 feet. The elevation of the lake's surface is 1495 feet above sea level. What is the elevation (with respect to sea level) of the deepest point in the lake? (*Source*: U.S. Geological Survey)

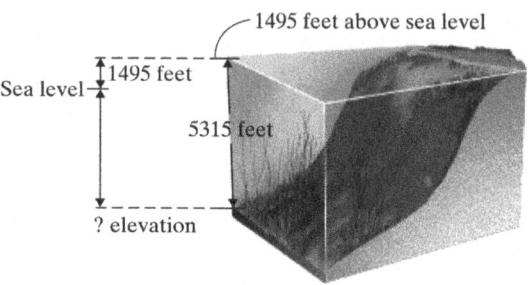

1495 feet above sea level

Sea level — 1495 feet

5315 feet

? elevation

30. Find the average of $-12, -13, 0, 9$.

31. Translate the following phrases into mathematical expressions. Use x to represent "a number."
   a. The product of a number and 17
   b. A number subtracted from 20

For Exercises 32 through 35, solve.

32. $-9n = -45$

33. $\dfrac{n}{-7} = 4$

34. $x - 16 = -36$

35. $-20 + 8 + 8 = x$

Section 3.1 Simplifying Algebraic Expressions

Objectives

A Use Properties of Numbers to Combine Like Terms
B Use Properties of Numbers to Multiply Expressions
C Simplify Expressions by Multiplying and Then Combining Like Terms
D Find the Perimeter and Area of Figures

Directions: Complete your Interactive Organizer by filling in the blanks and solving exercises as you complete each screen of the Interactive Assignment.

- For **WORK WITH ME** exercises, follow along and write each step needed and shown to solve, including the final answer.
- For **YOUR TURN** exercises, write the exercise generated for you in MyLab Math, then "show your work" by writing each step needed to solve, including the final answer.

Objective A: Use Properties of Numbers to Combine Like Terms

The addends of an algebraic expression are called the _____ of the expression.

How many terms are in $x + 3$? _____

How many terms are in $3y^2 + (-6y) + 4$? _____

A term that is only a number has a special name. It is called a _____ , or simply a _____ .

A term that contains a variable is called a _____ .

In the expression $x + 3$,

What type of term is x? _____   What type of term is 3? _____

In the expression $3y^2 + (-6y) + 4$,

What type of terms are $3y^2$ and $-6y$? _____ What type of term is 4? _____

The number factor of a variable term is called the _____ or simply the _____ .

A numerical coefficient of _____ is usually not written.

What is the numerical coefficient of $5x$? _____

What is the understood numerical coefficient of x?

What is the numerical coefficient of $3y^2$?   _____

What is the numerical coefficient of $-6y$?   _____

Terms with the same variable factors, except that they may have different numerical coefficients, are called _____ .

WORK WITH ME #1.

Fill in the table below.

Like Terms	Unlike Terms

| $3x, -4x$ | $7x, \ 7y$ | $-6y, \ 2y, \ y$ | $5x, \ x^2$ |

WORK WITH ME #2.

$7x + 2x$

WORK WITH ME #3.

Simplify each expression by combining like terms.

a. $3x + 5x$ b. $3a + 2a + 7a - 5$

YOUR TURN #1:

Helpful Hint

Remember that…

commutative involves a change in _____ : $a + b = b + a$

$$a \cdot b = b \cdot a$$

associative involves a change in _____ : $(a + b) + c = a + (b + c)$

$$(a \cdot b) \cdot c = a \cdot (b \cdot c)$$

WORK WITH ME #4.

Simplify: $2y - 6 + 4y + 8$

YOUR TURN #2:

Objective B: Use Properties of Numbers to Multiply Expressions.

Watch the objective video.

Addition and Multiplication are: Commutative (can change _____) and
Associative (can change _____)

VIDEO WORK WITH ME.

Multiply. $-3(11y)$ Multiply. $3(a - 6)$

YOUR TURN #1:   **YOUR TURN #2:**

YOUR TURN #3:   **YOUR TURN #4:**

Objective C: Simplify Expressions by Multiplying and Then Combining Like Terms.

WORK WITH ME #1.

Simplify: $2(3 + 7x) - 15$

WORK WITH ME #2.

Simplify: $-7(x + 5) + 5(2x + 1)$

YOUR TURN #1:

Objective D: Find the Perimeter and Area of Figures.

WORK WITH ME #1.

Find the perimeter of the triangle.

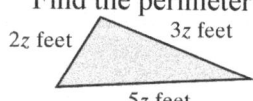

WORK WITH ME #2.

Find the area of this YMCA basketball court.

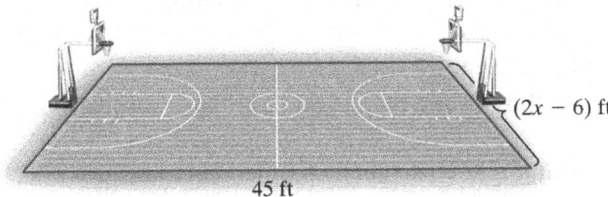

$(2x - 6)$ ft

45 ft

Helpful Hint

Don't forget.....

How is Area Measured?   How is Perimeter Measured?

YOUR TURN #1:   **YOUR TURN #2:**

Section 3.2 Solving Equations: Review of the Addition and Multiplication Properties

Objectives
A Use the Addition Property or the Multiplication Property to Solve Equations
B Use Both Properties to Solve Equations
C Translate Word Phrases to Mathematical Expressions

Directions: Complete your Interactive Organizer by filling in the blanks and solving exercises as you complete each screen of the Interactive Assignment.

- For **WORK WITH ME** exercises, follow along and write each step needed and shown to solve, including the final answer.
- For **YOUR TURN** exercises, write the exercise generated for you in MyLab Math, then "show your work" by writing each step needed to solve, including the final answer.

Objective A: Use the Addition Property or the Multiplication Property to Solve Equations

WORK WITH ME #1. Fill in the table below.

Equations	Expressions

$7x = 6x + 4$ $y - 1 + 11y - 21$ $7x - 6x + 4$ $3(3y - 5) = 10y$

WORK WITH ME #2.

The same number may be _____ to or _____ from both sides of an equation without changing the _____ of the equation.

Also, both sides of an equation may be _____ or _____ by the same nonzero number without changing the _____ of the equation.

WORK WITH ME #3.   **YOUR TURN #1:**

Solve: $3w - 12w = -27$

WORK WITH ME #4.   **YOUR TURN #2:**

Solve: $3(3x - 5) = 10x$

Objective B: Use Both Properties to Solve Equations

WORK WITH ME #1.   **YOUR TURN #1:**

Solve: $2x - 6 = 18$

WORK WITH ME #2.   **YOUR TURN #2:**

Solve: $-3(x + 9) - 41 = 4 - 60$

Objective C: Translate Word Phrases to Mathematical Expressions

WORK WITH ME #1. Fill in the table below.

Addition	Subtraction	Multiplication	Division
added to	subtracted from	multiply	shared equally among
more than	less than	twice	per
total	less	twice/double/triple	divided by
			divided into

Divided	Sum	Difference	Product	Times	Quotient	Plus	Minus

Increased by	Of	Decreased by

WORK WITH ME #2.

Write each phrase as an algebraic expression. Use x to represent "a number."

a. Eleven subtracted from a number

b. The product of -11 and a number, increased by 5

YOUR TURN #1:   **YOUR TURN #2:**

Section 3.3 Solving Linear Equations in One Variable

Objectives
>   A Solve Linear Equations Using the Addition and Multiplication Properties
>   B Solve Linear Equations Containing Parentheses
>   C Write Numerical Sentences as Equations

Directions: Complete your Interactive Organizer by filling in the blanks and solving exercises as you complete each screen of the Interactive Assignment.
- For **WORK WITH ME** exercises, follow along and write each step needed and shown to solve, including the final answer.
- For **YOUR TURN** exercises, write the exercise generated for you in MyLab Math, then "show your work" by writing each step needed to solve, including the final answer.

Objective A: Solve Linear Equations Using the Addition and Multiplication Properties

An equation such as $5x - 2 = 6x$ is a _____ .

It is called linear or *first-degree* for the following reasons:

1. _____

2. _____

3. _____

Helpful Hint

Solve $x + 2 = 10$ using the

Addition Property

Solve $2x = 10$ using the

Multiplication Property

WORK WITH ME #1.

   Solve: $10x + 15 = 6x + 3$

YOUR TURN #1:

WORK WITH ME #2.

   Solve: $17 - 7x + 3 = -3x + 21 - 3x$

YOUR TURN #2:

<u>Objective B: Solve Linear Equations Containing Parentheses</u>

Write the steps in your own words.

Steps for Solving an Equation
Step 1
Step 2
Step 3
Step 4
Step 5

WORK WITH ME #1.

Solve $3(5c - 1) - 2 = 13c + 3$

YOUR TURN #1:

Objective C: Write Numerical Sentences as Equations

WORK WITH ME #1. Circle the word in each sentence that means equals.

1. 3 equals 2 plus 1.

2. The quotient of 10 and –5 gives –2.

3. 17 minus 12 is 5.

4. 11 plus 2 yields 13.

5. Twice –15 amounts to –30.

6. –24 is equal to 2 times –12.

WORK WITH ME #2. Use the information in the boxes below to write the following sentences as equations.

| 7 | 6 | 2 | 26 | 3 | 42 | = | × | − | + | (|) |

The product of 7 and 6 is 42.

Twice the sum of 7 and 6 is equal to 26.

The quotient of 6 and 2 yields 3.

YOUR TURN #1:

YOUR TURN #2:

Section 3.4 Linear Equations in One Variable and Problem Solving

Objectives
A Writing Sentences as Equations
B Use Problem-Solving Steps to Solve Problems

Directions: Complete your Interactive Organizer by filling in the blanks and solving exercises as you complete each screen of the Interactive Assignment.

- For **WORK WITH ME** exercises, follow along and write each step needed and shown to solve, including the final answer.
- For **YOUR TURN** exercises, write the exercise generated for you in MyLab Math, then "show your work" by writing each step needed to solve, including the final answer.

<u>**Objective A: Writing Sentences as Equations**</u>

Watch the objective video.

Fill in the chart with key words and phrases that represent the operations and symbol shown.

Addition	Subtraction	Multiplication	Division	Equals Sign

***VIDEO* WORK WITH ME.**

Twice a number gives 108.   A number subtracted from −20 amounts to 104.

WORK WITH ME #1.

Translate to an equation: A number subtracted from 3 is 14.

YOUR TURN #1:   **YOUR TURN #2:**

Objective B: Use Problem-Solving Steps to Solve Problems.

Write the steps in your own words.

Problem-Solving Steps
Step 1
Step 2
Step 3
Step 4

WORK WITH ME #1.

Translate into an equation. Then solve the equation.

Three times a number, added to 9, is 33. Find the number.

YOUR TURN #1:

WORK WITH ME #2.

A falcon, when diving, can travel five times as fast as a pheasant's top speed. If the total speed for these two birds is 222 miles per hour, find the fastest speed of the falcon and the fasted speed of the pheasant. (*Source: Fantastic Book of Comparisons*)

YOUR TURN #2:

Chapter 3 Review and Practice

Study Skills
Chapter Vocabulary
Getting Ready for the Test
Review Exercises
Practice Chapter Test

Study Skills

Directions: **Watch the Study Skills video.**

Chapter Vocabulary

WORK WITH ME.

Fill in each blank with one of the words or phrases listed below:

| variable | addition | constant | algebraic expression | equation |

| terms | simplified | multiplication | evaluating the expression | solution |

| like | combined | numerical coefficient | distributive |

1. An algebraic expression is _____ when all the terms have been _____ .

2. Terms that are exactly the same, except that they may have been different numerical coefficients, are called _____ terms.

3. A letter used to represent a number is called a(n) _____ .

4. A combination of operations on variables and numbers is called a(n) _____ .

5. The addends of an algebraic expression are called the _____ of the expression.

6. The number factor of a variable terms is called the _____ .

7. Replacing a variable in an expression by a number and then finding the value of the expression is called _____ for the variable.

8. A term that is a number only is called a(n) _____ .

9. A(n) _____ is the form expression = expression.

10. A(n) _____ of an equation is a value for the variable that makes the equation a true statement.

11. To multiply $-3(2x + 1)$, we use the _____ property.

12. By the _____ property of equality, we may multiply or divide both sides of an equation by any nonzero number without changing the solution of the equation.

13. By the _____ property of equality, the same number may be added to or subtracted from both sides of an equation without changing the solution of the equation.

Getting Ready for the Test.

- These exercises will help you avoid common errors while taking your chapter test.

General Directions: Read the exercise Write any notes or steps in this Interactive Organizer, along with your answer to the exercise. In the MyLab Math Interactive Assignment, click the **SHOW ANSWERS** button to check your answers. Correct any errors, or press the **PLAY** button for a video solution.

Multiple Choice. Simplify each expression in Exercises 1 through 6. Then choose the simplified for: A, B, C, D, or E. Choices may be used more than once or not at all.

  A. $-6x$ B. $6x$ C. $4x + 2$ D. $4x + 1$ E. $4x + 6$

1. $x - 7x$

2. $-2x + 3 + 8x - 3$

3. $-3(2x)$

4. $2(2x + 1)$

5. $-(x + 2) + 5x + 4$

6. $3(1 + 2x) - 3$

Multiple Choice.

7. To solve $5x - 10 = 0$, we will first add 10 to each side of the equation. Once this is done, the equivalent equation is
    A. $5x = 0$ B. $5x = 10$ C. $5x = -10$

8. To solve $4(2x + 1) - 8 = 14x - 10$, we will first simplify the left side of the equation. Once this is done, the equivalent equation is
    A. $8x - 4 = 14x - 10$ B. $8x - 7 = 14x - 10$ C. $2x - 7 = 14x - 10$

Matching. Let x represent "a number." Match each phrase in the first column with its translated variable expression in the second column.

9. the sum of -3 and a number   A. $-3x$

10. the product of -3 and a number   B. $3 + x$

11. 5 subtracted from a number   C. $-3 + x$

12. 5 minus a number   D. $x - 5$

                                        E. $5 - x$

                                        F. $5x$

Matching. Let x represent "a number." Match each sentence in the first column with its translated equation in the second or third column.

13. Twice a number gives 14.   A. $6x = 14$ E. $x - 2 = 14$

14. A number divided by 6 yields 14.   B. $2x = 14$ F. $2 + 6x = 14$

15. Two subtracted from a number equals 14.   C. $\dfrac{x}{6} = 14$ G. $6(2 + x) = 14$

16. Six times a number, added to 2 is 14.   D. $2 - x = 14$

Review Exercises

In the **MyLab Math, Interactive Assignment, Review Exercises** section, there are algorithmically generated "Your Turn" exercises so that you can check your knowledge of some core concepts in this chapter. Insert a few sheets of paper in your Interactive Organizer to "record and show your work" along with the final answer.

Practice Chapter Test

- These exercises will help you practice for your chapter test.

General Directions: For each exercise, "show your work" by writing each step in the solution process within your Interactive Organizer, including your final answer. In the MyLab Math Interactive Assignment, click the Show Answer button to check your answer. Correct any errors, or press the **PLAY** button for a video solution.

1. Simplify $7x - 5 - 12x + 10$ by combining like terms.

2. Multiply: $-2(3y + 7)$

3. Simplify: $-(3z + 2) - 5z - 18$

4. Write an expression that represents the Perimeter of the equilateral triangle (a triangle with three sides of equal length.) Simplify the expression.

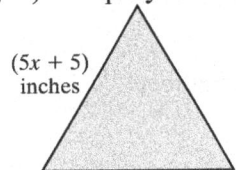

$(5x + 5)$ inches

5. Write an expression that represents the area of the rectangle. Simplify the expression.

4 meters

Rectangle $(3x - 1)$ meters

Solve each equation.

6. $12 = y - 3y$

7. $\dfrac{x}{2} = -5 - (-2)$

8. $5x + 12 - 4x - 14 = 22$

9. $-4x + 7 = 15$

10. $2(x-6)=0$

11. $-4(x-11)-34=10-12$

12. $5x-2=x-10$

13. $4(5x+3)=2(7x+6)$

14. $6+2(3x-1)=28$

Translate the following phrases into mathematical expressions. If needed, use x to represent "a number."

15. The sum of −23 and a number

16. Three times a number, subtracted from −2

Translate each sentence into an equation. If needed, use x to represent "a number."

17. The sum of twice 5 and −15 is −5.

18. Six added to three times a number equals −30.

Solve.

19. The difference of three times a number and five times the same number is 4. Find the number.

20. In a championship basketball game, Paula made twice as many free throws as Maria. If the total number of free throws made by both women was 12, find how many free throws Paula made.

21. In a 10-kilometer race, there are 112 more men entered than women. Find the number of female runners if the total number of runners in the race is 600.

Section 4.1 Introduction to Fractions and Mixed Numbers

Objectives
      A Identify the Numerator and the Denominator of a Fraction
      B Write a Fraction to Represent Parts of Figures or Real-Life Data
      C Graph Fractions on a Number Line
      D Review Division Properties of 0 and 1
      E Write Mixed Numbers as Improper Fractions
      F Write Improper Fractions as Mixed Numbers or Whole Numbers

Directions: Complete your Interactive Organizer by filling in the blanks and solving exercises as you complete each screen of the Interactive Assignment.
- For **WORK WITH ME** exercises, follow along and write each step needed and shown to solve, including the final answer.
- For **YOUR TURN** exercises, write the exercise generated for you in MyLab Math, then "show your work" by writing each step needed to solve, including the final answer.

Objective A: Identify the Numerator and the Denominator of a Fraction

Watch the objective video.

We can use _____ to represent part of a whole.

$$\frac{1}{2} \qquad \frac{10}{3}$$

The number on top is the _____ .

The number on bottom is the _____ .

The bar between the numbers is the _____ bar.

When the numerator is less than the denominator, it is a _____ fraction.

When the numerator is greater than or equal to the denominator, it is a _____ fraction.

YOUR TURN #1:   **YOUR TURN #2:**

Objective B: Write a Fraction to Represent Parts of Figures or Real-Life Data

Write an improper fraction to represent the shaded portion of the diagram.

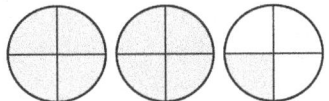

A _____ **number** contains a whole number and a fraction. Mixed numbers are greater than 1.

The shaded part of the group of circles below is an improper fraction $\frac{9}{4}$. Now let's write the shaded part as a mixed number.

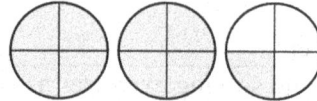

Thus, $\frac{9}{4} =$

Helpful Hint
The mixed number, _____ , diagrammed above represents $2 +$ _____ .

WORK WITH ME #1.

A bag contains 50 red or blue marbles. If 21 marbles are blue, answer each question.
  a. What fraction of the marbles are blue?   b. How many marbles are red?

  c. What fraction of the marbles are red?

YOUR TURN #1:   **YOUR TURN #2:**

Objective C: Graph Fractions on a Number Line

Watch the objective video.

$$\frac{1}{4}$$

$$\frac{7}{3}$$

YOUR TURN #1:   **YOUR TURN #2:**

Objective D: Review Division Properties of 0 and 1

Watch the objective video.

$$\frac{-5}{1} \qquad\qquad\qquad \frac{3}{1} \qquad\qquad\qquad \frac{-8}{-8}$$

$$\frac{0}{-2} \qquad\qquad\qquad \frac{-9}{0}$$

Let *n* be any integer except 0.

$$\frac{n}{n} = \qquad\qquad \frac{0}{n} = \qquad\qquad \frac{n}{1} = \qquad\qquad \frac{n}{0} \text{ is } \underline{\hspace{2cm}} .$$

YOUR TURN #1:   **YOUR TURN #2:**

YOUR TURN #3:   **YOUR TURN #4:**

Objective E: Write Mixed Numbers as Improper Fractions

Mixed numbers and improper fractions can both be used to represent the shaded part of figure groups.

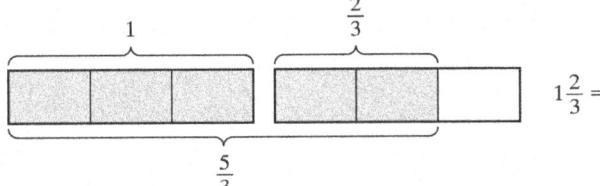

$$1\tfrac{2}{3} =$$

Writing a Mixed Number as an Improper Fraction

Step 1

Step 2

Step 3

WORK WITH ME #1.

Write each mixed number as an improper fraction.

a. $2\dfrac{1}{3}$

b. $9\dfrac{7}{20}$

YOUR TURN #1:   **YOUR TURN #2:**

Objective F: Write Improper Fractions as Mixed Numbers or Whole Numbers

Watch the objective video.

The fraction bar means _____ .

Writing an Improper Fraction as a Mixed Number or a Whole Number

Step 1:

Step 2:

$\dfrac{17}{5}$ $\dfrac{37}{8}$

YOUR TURN #1:   **YOUR TURN #2:**

Section 4.2 Factors and Simplest Form

Objectives

  A Write a Number as a Product of Prime Numbers
  B Write a Fraction in Simplest Form
  C Determine Whether Two Fractions Are Equivalent
  D Solve Problems by Writing Fractions in Simplest Form

Directions: Complete your Interactive Organizer by filling in the blanks and solving exercises as you complete each screen of the Interactive Assignment.

- For **WORK WITH ME** exercises, follow along and write each step needed and shown to solve, including the final answer.
- For **YOUR TURN** exercises, write the exercise generated for you in MyLab Math, then "show your work" by writing each step needed to solve, including the final answer.

Objective A: Write a Number as a Product of Prime Numbers

Watch the objective video.

A _____ **number** is a natural number greater than 1 whose only factors are 1 and itself. The first few _____ numbers are 2, 3, 5, 7, 11, 13, 17, 19, 23, 29, …

A _____ **number** is a natural number greater than 1 that is not prime.

VIDEO WORK WITH ME.

20 =

When a composite number is written as a product of prime numbers, this product is called the _____ factorization of the number.

There is only _____ prime factorization for each composite number.

VIDEO WORK WITH ME.

48   240

Below are a few quick _____ **tests** to determine whether a number is divisible by the primes 2, 3, or 5.

A number is divisible by 2, for example, if 2 divides it _____ so that the remainder is _____.

Divisibility Tests
      A whole number is divisible by

Divisibility by 2

Divisibility by 3

Divisibility by 5

YOUR TURN #1:   **YOUR TURN #2:**

Objective B: Write a Fraction in Simplest Form

Fractions that represent the same portion of a whole or the same point on a number line are called _____ **fractions**.

Equivalent Fractions

Figures	**Number Line**
Shade $\frac{1}{3}$ and $\frac{2}{6}$ on the same-sized figures,	Graph $\frac{1}{3}$ and $\frac{2}{6}$ on a number line,

Both $\frac{1}{3}$ and $\frac{2}{6}$ represent the same portion of a whole. These fractions are called **equivalent fractions;** $\frac{1}{3} =$

Both $\frac{1}{3}$ and $\frac{2}{6}$ correspond to the same point. These fractions are called **equivalent fractions;** $\frac{1}{3} =$

A special equivalent form of a fraction is called _____ **form.**

Simplest Form of a Fraction
A fraction is written in **simplest form** or _____ **terms** when the numerator and the
denominator have _____ common factors other than 1.

Equivalent Fractions $\dfrac{8}{12} =$

The process of writing a fraction in simplest form is called _____ the fraction.

To simplify a fraction, we use the Fundamental Property of Fractions.

$\dfrac{a \cdot c}{b \cdot c} =$

The Fundamental Property of Fractions allows us to simplify $\dfrac{12}{20}$, for example.

Notice that 12 and 20 have a common factor of _____ .

$\dfrac{12}{20} =$

We can use a shortcut procedure with common factors when simplifying.

$\dfrac{4}{6} =$

Why does this work?
    This procedure is possible because dividing out a common _____ in the numerator and
    denominator is the same as removing a factor of _____ in the product.

To write a fraction in simplest form, use the **Fundamental Property of Fractions** and _____
the numerator and denominator by any common factors.

Helpful Hint
Writing the _____ factorizations of the numerator and the denominator is helpful in finding
any common factors.

WORK WITH ME #1.

Write each fraction in simplest form.

a. $-\dfrac{63}{81}$

b. $\dfrac{30x^2}{36x}$

YOUR TURN #1:   **YOUR TURN #2:**

YOUR TURN #3:

Objective C: Determine Whether Two Fractions Are Equivalent

Watch the objective video.

***VIDEO* WORK WITH ME.**

$$\frac{7}{11} \overset{?}{=} \frac{5}{8}$$ $$\frac{3}{9} \overset{?}{=} \frac{6}{18}$$

YOUR TURN #1:   **YOUR TURN #2:**

Objective D: Solve Problems by Writing Fractions in Simplest Form

Watch the objective video.

***VIDEO* WORK WITH ME.**

The outer wall of the Pentagon is 24 inches wide. Ten inches is concrete, 8 inches is brick, and 6 inches is limestone. What fraction of the wall is concrete?

YOUR TURN #1:

Section 4.3 Multiplying and Dividing Fractions

Objectives
    A Multiply Fractions
    B Evaluate Exponential Expressions with Fractional Bases
    C Divide Fractions
    D Multiply and Divide Given Fractional Replacement Values
    E Solve Applications That Require Multiplication of Fractions

Directions: Complete your Interactive Organizer by filling in the blanks and solving exercises as you complete each screen of the Interactive Assignment.
- For **WORK WITH ME** exercises, follow along and write each step needed and shown to solve, including the final answer.
- For **YOUR TURN** exercises, write the exercise generated for you in MyLab Math, then "show your work" by writing each step needed to solve, including the final answer.

Objective A: Multiply Fractions

Watch the objective video.

Multiplying Fractions
To multiply two fractions, multiply the _____ and multiply the _____ .
  If a, b, c, and d represent numbers and b and d are not 0, we have

$$\frac{a}{b} \cdot \frac{c}{d} =$$

***VIDEO* WORK WITH ME.**

$$\frac{2}{3} \cdot \frac{5}{9}$$
                            $$-\frac{2}{7} \cdot \frac{5}{8}$$
                                           $$\frac{x^2}{y} \cdot \frac{y^3}{x}$$

YOUR TURN #1:
                                            YOUR TURN #2:

YOUR TURN #3:

Objective B: Evaluate Exponential Expressions with Fractional Bases

Watch the objective video.

***VIDEO* WORK WITH ME.**

$$\left(-\frac{2}{3}\right)^2$$

WORK WITH ME #1.   **YOUR TURN #1:**

Objective C: Divide Fractions

Before we can divide fractions, we need to know how to find the _____ of a fraction.

Reciprocal of a Fraction
Two numbers are **reciprocals** of each other if their product is _____ . The reciprocal of the fraction

$\frac{a}{b}$ is $\frac{b}{a}$

because $\frac{a}{b} \cdot \frac{b}{a} =$

Helpful Hint
Every number has a reciprocal except _____ . The number 0 has _____ reciprocal because there is no number such that $0 \cdot a = 1$.

Example   The reciprocal of $\frac{2}{5}$ is _____ because $\frac{2}{5} \cdot \text{——} = \text{——} = 1$.

Another Example   The reciprocal of 5 is _____ because $5 \cdot \text{——} = \text{——} \cdot \text{——} = \text{——} = 1$.

Third Example   The reciprocal of $-\frac{7}{11}$ is _____ because $-\frac{7}{11} \cdot \text{——} = \text{——} = 1$.

Dividing Fractions

To divide two fractions, multiply the first fraction by the _____ of the second fraction.

If a, b, c, and d represent numbers, and b, c, and d are not 0, then

$$\frac{a}{b} \div \frac{c}{d} =$$

WORK WITH ME #1.

Divide fractions.

$$\frac{3}{4} \div \frac{1}{8}$$

WORK WITH ME #2.

a. $\dfrac{4}{8} \div \dfrac{3}{16}$

b. $\left(\dfrac{1}{2} \cdot \dfrac{2}{3} \right) \div \dfrac{5}{6}$

YOUR TURN #1:   **YOUR TURN #2:**

YOUR TURN #3:

Objective D: Multiply and Divide Given Fractional Replacement Values

Watch the objective video.

VIDEO WORK WITH ME.

If $x = -\dfrac{4}{5}$ and $y = \dfrac{9}{11}$, find

     a) xy b) $x \div y$.

WORK WITH ME #1.   **YOUR TURN #1:**

Objective E: Solve Applications That Require Multiplication of Fractions

Watch the objective video.

***VIDEO* WORK WITH ME.**

The radius of a circle is one-half of its diameter. If the diameter of a circle is $\frac{3}{8}$ of an inch, what is the radius?

YOUR TURN #1:   **YOUR TURN #2:**

Section 4.4 Adding and Subtracting Like Fractions, Least Common Denominator, and Equivalent Fractions

Objectives
    A Add or Subtract Like Fractions
    B Add or Subtract Given Fractional Replacement Values
    C Solve Problems by Adding or Subtracting Like Fractions
    D Find the Least Common Denominator of a List of Fractions
    E Write Equivalent Fractions

Directions: Complete your Interactive Organizer by filling in the blanks and solving exercises as you complete each screen of the Interactive Assignment.
  - For **WORK WITH ME** exercises, follow along and write each step needed and shown to solve, including the final answer.
  - For **YOUR TURN** exercises, write the exercise generated for you in MyLab Math, then "show your work" by writing each step needed to solve, including the final answer.

Objective A: Add or Subtract Like Fractions

Fractions with the same denominator are called _____ **fractions**.

Fractions that have different denominators are called _____ **fractions**.

Like Fractions	**Unlike Fractions**
$\dfrac{2}{5}$ and	$\dfrac{2}{5}$ and
$\dfrac{5}{21}$, $\dfrac{16}{21}$, and	$\dfrac{5}{7}$ and

To see how we add like fractions, study one or both illustrations below.

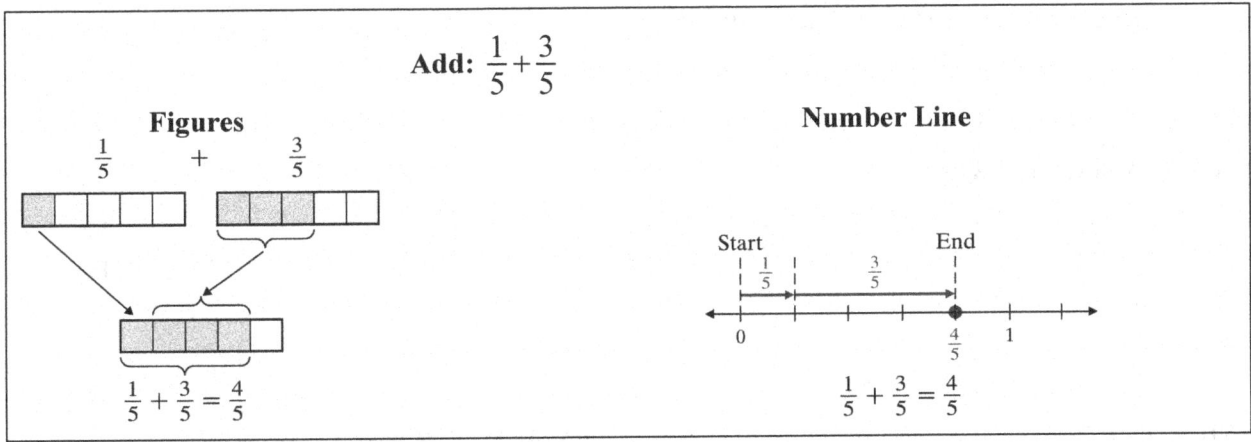

$$\text{Add: } \frac{1}{5} + \frac{3}{5}$$

81

Adding or Subtracting Like Fractions (Fractions with the Same Denominator)
If a, b, and c are numbers and b is not 0, then

$$\frac{a}{b} + \frac{c}{b} = \qquad \text{and also} \quad \frac{a}{b} - \frac{c}{b} =$$

In other words, to add or subtract fractions with the same denominator, add or subtract their numerators and write the sum or difference over the _____ denominator.

$$\frac{1}{4} + \frac{2}{4} = \frac{}{4} = \frac{}{4} \quad \begin{array}{l}\text{Add the numerators.}\\ \text{Keep the denominator.}\end{array} \qquad \frac{4}{5} - \frac{2}{5} = \frac{}{5} = \frac{}{5} \quad \begin{array}{l}\text{Subtract the numerators.}\\ \text{Keep the denominator.}\end{array}$$

Helpful Hint   Don't forget to write all answers in _____ form.

WORK WITH ME #1.

Add or subtract, then simplify.

a. $\dfrac{5}{11} + \dfrac{2}{11}$

b. $\dfrac{10}{11} - \dfrac{4}{11}$

WORK WITH ME #2.

We know that $\dfrac{-12}{6} = \dfrac{12}{-6} = -\dfrac{12}{6}$ since these all simplify to -2.

   Study the above. What is true in general?

$$\frac{}{b} = \frac{a}{} = -\frac{a}{b} \quad \text{as long as } b \text{ is not 0.}$$

WORK WITH ME #3.

a. $-\dfrac{6}{20} + \dfrac{1}{20}$

b. $\dfrac{7x}{16} - \dfrac{15x}{16}$

c. $\dfrac{9}{12} - \dfrac{7}{12} - \dfrac{10}{12}$

YOUR TURN #1:

YOUR TURN #2:

YOUR TURN #3:

Objective B: Add or Subtract Given Fractional Replacement Values

Watch the objective video.

***VIDEO* WORK WITH ME.**

Evaluate $x - y$ if $x = -\dfrac{1}{5}$ and $y = -\dfrac{3}{5}$.

YOUR TURN #1:

Objective C: Solve Problems by Adding or Subtracting Like Fractions

Watch the objective video.

Perimeter means _____ around.

***VIDEO* WORK WITH ME.**

$\dfrac{5}{12}$
meters

$\dfrac{7}{12}$ meters

YOUR TURN #1:

Objective D: Find the Least Common Denominator of a List of Fractions

To add and subtract fractions that have different, or unlike, denominators, we first write them as _____ fractions with a _____ denominator.

Although any common denominator can be used, we will use the _____ **common denominator (LCD).**

Definition of the LCD
The **least common denominator (LCD)** of a list of fractions is the _____ positive number divisible by all the denominators in the list. (The least common denominator is also the **least common _____ (LCM) of the denominators.)**

Finding the LCD: Method 1

One way to find the LCD is to check _____ multiples of the larger denominator until the LCD is found.

Method 1: Finding the LCD of a List of Fractions Using Multiples of the Largest Number

Step 1

Step 2

WORK WITH ME #1.

Find the LCD of $\dfrac{2}{9}$ and $\dfrac{6}{15}$.

Finding the LCD: Method 2

A second method for finding the LCD uses _____ factorization.

Method 2: Finding the LCD of a List of Denominators Using Prime Factorization

Step 1

Step 2

Step 3

WORK WITH ME #2.

Find the LCD of $\dfrac{4}{3}$, $\dfrac{8}{21}$, and $\dfrac{3}{56}$.

YOUR TURN #1:   **YOUR TURN #2:**

Objective E: Write Equivalent Fractions

Watch the objective video.

***VIDEO* WORK WITH ME.**

$\dfrac{2}{3} = \dfrac{}{21}$ $\dfrac{2y}{3} = \dfrac{}{12}$

$\dfrac{5}{9} = \dfrac{}{36a}$

YOUR TURN #1:   **YOUR TURN #2:**

YOUR TURN #3:

Section 4.5 Adding and Subtracting Unlike Fractions

Objectives
A Add or Subtract Unlike Fractions
B Write Fractions in Order
C Evaluate Expressions Given Fractional Replacement Values
D Solve Problems by Adding or Subtracting Unlike Fractions

Directions: Complete your Interactive Organizer by filling in the blanks and solving exercises as you complete each screen of the Interactive Assignment.

- For **WORK WITH ME** exercises, follow along and write each step needed and shown to solve, including the final answer.
- For **YOUR TURN** exercises, write the exercise generated for you in MyLab Math, then "show your work" by writing each step needed to solve, including the final answer.

Objective A: Add or Subtract Unlike Fractions

To add or subtract unlike fractions, we first rewrite the fractions as _____ fractions with a _____ denominator and then add or subtract the like fractions.

The common denominator we will use is called the least _____ denominator (LCD).

Adding or Subtracting Unlike Fractions
Step 1
Step 2
Step 3
Step 4

WORK WITH ME #1.

$\dfrac{7}{8} - \dfrac{5}{6}$

WORK WITH ME #2.

$\dfrac{1}{2} + \dfrac{2}{3} + \dfrac{5}{6}$

WORK WITH ME #3.

Add or subtract as indicated.

a. $-\dfrac{9}{12} + \dfrac{17}{24} - \dfrac{1}{6}$

b. $\dfrac{5}{9} + \dfrac{1}{y}$

YOUR TURN #1:   **YOUR TURN #2:**

YOUR TURN #3:   **YOUR TURN #4:**

Objective B: Write Fractions in Order

Watch the objective video.

VIDEO WORK WITH ME.

Insert < or > to form a true statement.

$$\frac{2}{7} \quad \frac{3}{10}$$

YOUR TURN #1:   **YOUR TURN #2:**

Objective C: Evaluate Expressions Given Fractional Replacement Values

Watch the objective video.

VIDEO WORK WITH ME.

Evaluate $2y + x$ if $x = \frac{1}{3}$ and $y = \frac{3}{4}$.

YOUR TURN #1:

Objective D: Solve Problems by Adding or Subtracting Unlike Fractions

Watch the objective video.

***VIDEO* WORK WITH ME.**

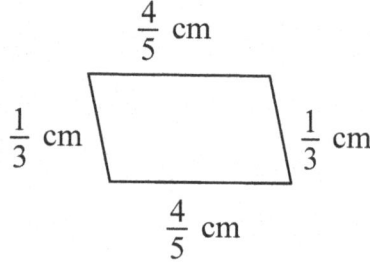

WORK WITH ME #1.

Find the difference in length of two boards if one board is $\frac{4}{5}$ foot long and the other is $\frac{2}{3}$ foot long.

YOUR TURN #1:

Section 4.6 Complex Fractions and Review of Order of Operations

Objectives
    A Simplify Complex Fractions
    B Review the Order of Operations
    C Evaluate Expressions Given Replacement Values

Directions: Complete your Interactive Organizer by filling in the blanks and solving exercises as you complete each screen of the Interactive Assignment.
- For **WORK WITH ME** exercises, follow along and write each step needed and shown to solve, including the final answer.
- For **YOUR TURN** exercises, write the exercise generated for you in MyLab Math, then "show your work" by writing each step needed to solve, including the final answer.

Objective A: Simplify Complex Fractions

We now practice _____ fractions whose numerators or denominators themselves contain fractions. These fractions are called _____ fractions.

Complex Fraction
A fraction whose numerator or denominator or both numerator and denominator contain fractions is called a _____ **fraction**.

Examples

Method 1 for Simplifying Complex Fractions
Our first method for simplifying complex fractions makes use of the fact that a fraction bar means _____ .

WORK WITH ME #1.

Simplify the complex fraction.

$$\dfrac{\dfrac{1}{8}}{\dfrac{3}{4}}$$

YOUR TURN #1:

Method 2 for Simplifying Complex Fractions
Our second method for simplifying complex fractions is to multiply the numerator and the denominator of the complex fraction by the _____ of all the fractions in its numerator and its denominator.

WORK WITH ME #2.

Simplify the complex fraction.

$$\frac{\dfrac{3x}{4}}{5 - \dfrac{1}{8}}$$

YOUR TURN #2:

Objective B: Review the Order of Operations

Watch the objective video.

Order of Operations
Step 1:
Step 2:
Step 3:
Step 4:

***VIDEO* WORK WITH ME.**

$$\left(\frac{2}{9} + \frac{4}{9}\right)\left(\frac{1}{3} - \frac{9}{10}\right)$$

YOUR TURN #1:   **YOUR TURN #2:**

Objective C: Evaluate Expressions Given Replacement Values

Watch the objective video.

***VIDEO* WORK WITH ME.**

Evaluate $x^2 - yz$ if $x = -\dfrac{1}{3}$, $y = \dfrac{2}{5}$, and $z = \dfrac{5}{6}$.

YOUR TURN #1:   **YOUR TURN #2:**

Section 4.7 Operations on Mixed Numbers

Objectives
A Graph Positive and Negative Fractions and Mixed Numbers
B Multiply or Divide Mixed Numbers or Whole Numbers
C Add or Subtract Mixed Numbers
D Solve Problems Containing Mixed Numbers
E Perform Operations on Negative Mixed Numbers

Directions: Complete your Interactive Organizer by filling in the blanks and solving exercises as you complete each screen of the Interactive Assignment.

- For **WORK WITH ME** exercises, follow along and write each step needed and shown to solve, including the final answer.
- For **YOUR TURN** exercises, write the exercise generated for you in MyLab Math, then "show your work" by writing each step needed to solve, including the final answer.

Objective A: Graph Positive and Negative Fractions and Mixed Numbers

Watch the objective video.

VIDEO WORK WITH ME.

Graph: $4, \ \dfrac{1}{3}, \ -3, \ -3\dfrac{4}{5}, \ 1\dfrac{1}{3}$

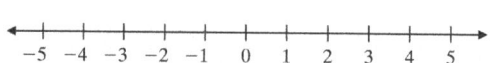

YOUR TURN #1:

Objective B: Multiply or Divide Mixed Numbers or Whole Numbers

Watch the objective video.

Multiplying or Dividing Fractions and Mixed Numbers or Whole Numbers
To multiply or divide with mixed number or whole numbers, first write any mixed or whole numbers as _____ fractions and then multiply or divide as usual.

VIDEO WORK WITH ME.

$3\dfrac{2}{3} \cdot 1\dfrac{1}{2}$ $7 \div 1\dfrac{3}{5}$

YOUR TURN #1:   **YOUR TURN #2:**

Objective C: Add or Subtract Mixed Numbers

Watch the objective video.

Adding or Subtracting Mixed Numbers To add or subtract mixed numbers, add or subtract the _____ parts and then add or subtract the _____ number parts.

VIDEO WORK WITH ME.

$$15\dfrac{4}{7}$$
$$+\ 9\dfrac{11}{14}$$

YOUR TURN #1:   **YOUR TURN #2:**

Objective D: Solve Problems Containing Mixed Numbers

WORK WITH ME #1.

Two rainbow trout weigh $2\dfrac{1}{2}$ pounds and $3\dfrac{2}{3}$ pounds. What is the total weight of the two trout?

YOUR TURN #1:   **YOUR TURN #2:**

Objective E: Perform Operations on Negative Mixed Numbers

Watch the objective video.

VIDEO WORK WITH ME.

$$-31\frac{2}{15}+17\frac{3}{20}$$

YOUR TURN #1:   **YOUR TURN #2:**

Section 4.8 Solving Equations Containing Fractions

Objectives

      A Solve Equations Containing Fractions
      B Solve Equations by Multiplying by the LCD
      C Review Adding and Subtracting Fractions

Directions: Complete your Interactive Organizer by filling in the blanks and solving exercises as you complete each screen of the Interactive Assignment.

- For **WORK WITH ME** exercises, follow along and write each step needed and shown to solve, including the final answer.
- For **YOUR TURN** exercises, write the exercise generated for you in MyLab Math, then "show your work" by writing each step needed to solve, including the final answer.

Objective A: Solve Equations Containing Fractions

Watch the objective video.

VIDEO WORK WITH ME.

$$x - \frac{1}{12} = \frac{5}{6}$$

Addition Property of Equality: The _____ number may be added or subtracted from both sides of an equation.

To check, replace the variable (in the _____ equation) with the proposed solution and see that a _____ statement results.

VIDEO WORK WITH ME.

$$-\frac{4}{9}z = -\frac{3}{2}$$

Two numbers are _____ of each other if their product is 1.

Multiplication Property of Equality: The _____ nonzero number may be multiplied to or divided by both sides of an equation.

YOUR TURN #1:   **YOUR TURN #2:**

Objective B: Solve Equations by Multiplying by the LCD

If an equation contains fractions, it is often helpful to first multiply both sides of the equation by the _____ of the fractions. This has the effect of _____ the fractions in the equation.

Let's review the steps for solving equations in x. An extra step is now included to handle equations containing fractions.

Solving an Equation in x

Step 1

Step 2

Step 3

Step 4

Step 5

Step 6

WORK WITH ME #1.

Solve.

$$\frac{7}{6}x = \frac{1}{4} - \frac{2}{3}$$

WORK WITH ME #2.

Solve.

$$\frac{z}{5} - \frac{z}{3} = 6$$

YOUR TURN #1:   **YOUR TURN #2:**

Objective C: Review Adding and Subtracting Fractions

Watch the objective video.

VIDEO WORK WITH ME.

$\dfrac{3x}{10} + \dfrac{x}{6}$

YOUR TURN #1:

Chapter 4 Review and Practice

> Study Skills
> Chapter Vocabulary
> Getting Ready for the Test
> Review Exercises
> Practice Chapter Test

Study Skills

Directions: **Watch the Study Skills video.**

Chapter Vocabulary

WORK WITH ME.

Fill in each blank with one of the words or phrases listed below:

| mixed number | like | numerator | prime factorization | composite | equivalent |

| cross products | denominator | prime number | improper fraction | simplest form |

| undefined | 0 | reciprocals | proper fraction |

1. Two numbers are _____ of each other if their product is 1.

2. A(n) _____ is a natural number greater than 1 that is not prime.

3. Fraction that represent the same portion of a whole are called _____ fractions.

4. A(n) _____ is a fraction whose numerator is greater than or equal to its denominator.

5. A(n) _____ is a natural number greater than 1 whose only factors are 1 and itself.

6. A fraction is in _____ when the numerator and the denominator have no factors in common other than 1.

7. A(n) _____ is one whose numerator is less than its denominator.

8. A(n) _____ contains a whole number part and a fraction part.

9. In the fraction $\frac{7}{9}$, the 7 is called the _____ and the 9 is called the _____.

10. The _____ of a number is the factorization in which all the factors are prime numbers.

11. The fraction $\frac{3}{0}$ is _____ .

12. The fraction $\frac{0}{5}$ = _____ .

13. Fractions that have the same denominator are called _____ fractions.

14. In $\frac{a}{b} = \frac{c}{d}$, $a \cdot d$ and $b \cdot c$ are called _____ .

Getting Ready for the Test.

- These exercises will help you avoid common errors while taking your chapter test.

General Directions: Read the exercise Write any notes or steps in this Interactive Organizer, along with your answer to the exercise. In the MyLab Math Interactive Assignment, click the **SHOW ANSWERS** button to check your answers. Correct any errors, or press the **PLAY** button for a video solution.

Multiple Choice. Exercises 1–10 are Multiple Choice. Choose the correct answer.

For Exercises 1 through 4, choose whether the expression simplifies to
   A. 1   B. −1   C. 0   D. undefined

1. $\frac{-2}{-2}$ 2. $\frac{-2}{2}$ 3. $\frac{2}{0}$ 4. $\frac{0}{-2}$

5. The mixed number $4\frac{3}{5}$ written as a fraction is:

   A. $\frac{12}{5}$ B. $\frac{5}{23}$ C. $\frac{23}{5}$ D. $\frac{23}{3}$

6. The improper fraction $\frac{23}{8}$ written as a mixed number is:

   A. 2.3   B. $2\frac{7}{8}$ C. $8\frac{2}{3}$ D. $2\frac{8}{7}$

For Exercises 7 through 10, the exercise statement and correct answer are given. Choose whether the correct operation in the box should be:
   A. + (addition)   B. − (subtraction)   C. · (multiplication)   D. ÷ (division)

7. $\frac{8}{11}\square\frac{2}{11}$; Answer: $\frac{16}{121}$ 8. $\frac{8}{11}\square\frac{2}{11}$; Answer: $\frac{8}{2}$ or 4

9. $\frac{8}{11}\square\frac{2}{11}$; Answer: $\frac{6}{11}$ 10. 7. $\frac{8}{11}\square\frac{2}{11}$; Answer: $\frac{10}{11}$

Matching. For Exercises 11 through 14, **Match** each operation of fraction s in the first column with the correct answer in the second or third column.

11. $\dfrac{5}{7} + \dfrac{1}{7}$

12. $\dfrac{5}{7} \cdot \dfrac{1}{7}$

13. $\dfrac{5}{7} \div \dfrac{1}{7}$

14. $\dfrac{5}{7} - \dfrac{1}{7}$

A. $\dfrac{5}{7}$

B. $\dfrac{5}{1}$ or 5

C. $\dfrac{6}{14}$ or $\dfrac{3}{7}$

D. $\dfrac{4}{7}$

E. undefined

F. $\dfrac{6}{7}$

G. $\dfrac{6}{49}$

H. $\dfrac{5}{49}$

Multiple Choice. Exercises 15–21 are Multiple Choice. Choose the correct answer.

For each expression in Exercises 15 through 18, use order of operations and decide which operation below should be performed first to simplify.

A. Addition   B. Subtraction   C. Multiplication   D. Division

15. $\dfrac{1}{2} + \dfrac{1}{5} \cdot \dfrac{1}{4}$

16. $\left(\dfrac{1}{2} + \dfrac{1}{5}\right) \cdot \dfrac{1}{4}$

17. $\dfrac{1}{2} \div \dfrac{1}{5} \cdot \dfrac{1}{4}$

18. $\dfrac{1}{2} \div \left(\dfrac{1}{5} \cdot \dfrac{1}{4}\right)$

19. To solve $\dfrac{x}{10} + \dfrac{3}{5} = 2$ let's multiply both sides of the equation by 10. Once this is done, the equivalent equation is:

   A. $x + 6 = 2$ B. $x + 3 = 2$ C. $x + 6 = 12$ D. $x + 6 = 20$

Equations are solved and expressions are simplified. For Exercises 20 through 21, decide whether each exercise is an

   A. expression   or an   B. equation

20. $\dfrac{2}{3} - \dfrac{1}{9}$

21. $\dfrac{2}{3} - \dfrac{1}{9} = \dfrac{x}{9}$

Review Exercises

In the **MyLab Math, Interactive Assignment, Review Exercises** section, there are algorithmically generated "Your Turn" exercises so that you can check your knowledge of some core concepts in this chapter. Insert a few sheets of paper in your Interactive Organizer to "record and show your work" along with the final answer.

Practice Chapter Test

- These exercises will help you practice for your chapter test.

General Directions: For each exercise, "show your work" by writing each step in the solution process within your Interactive Organizer, including your final answer. In the MyLab Math Interactive Assignment, click the Show Answer button to check your answer. Correct any errors, or press the **PLAY** button for a video solution.

Write a fraction to represent the shaded area.

1.

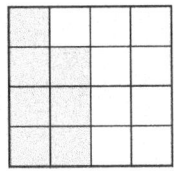

Write the mixed number as an improper fraction.

2. $7\dfrac{2}{3}$

Write the improper fraction as a mixed number.

3. $\dfrac{75}{4}$

Write each fraction in simplest form.

4. $\dfrac{24}{210}$ 5. $-\dfrac{42x}{70}$

Determine whether these fractions are equivalent.

6. $\dfrac{5}{7}$ and $\dfrac{8}{11}$ 7. $\dfrac{6}{27}$ and $\dfrac{14}{63}$

Find the prime factorization of each number.

8. 84   9. 495

Perform each indicated operation and write answers in simplest form.

10. $\dfrac{4}{4} \div \dfrac{3}{4}$ 11. $-\dfrac{4}{3} \cdot \dfrac{4}{4}$ 12. $\dfrac{7x}{9} + \dfrac{x}{9}$ 13. $\dfrac{1}{7} - \dfrac{3}{x}$

14. $\dfrac{xy^3}{z} \cdot \dfrac{z}{xy}$ 15. $-\dfrac{2}{3} \cdot -\dfrac{8}{15}$ 16. $\dfrac{9a}{10} + \dfrac{2}{5}$ 17. $-\dfrac{8}{15y} - \dfrac{2}{15y}$

18. $\dfrac{3a}{8} \cdot \dfrac{16}{6a^3}$

19. $\dfrac{11}{12} - \dfrac{3}{8} + \dfrac{5}{24}$

20. $\begin{aligned} 3\dfrac{7}{8} \\ 7\dfrac{2}{5} \\ +2\dfrac{3}{4} \\ \hline \end{aligned}$

21. $\begin{aligned} 19 \\ -2\dfrac{3}{11} \\ \hline \end{aligned}$

22. $-\dfrac{16}{3} \div -\dfrac{3}{12}$

23. $3\dfrac{1}{3} \cdot 6\dfrac{3}{4}$

24. $-\dfrac{2}{7}\left(6 - \dfrac{1}{6}\right)$

25. $\dfrac{1}{2} \div \dfrac{2}{3} \cdot \dfrac{3}{4}$

26. $\left(-\dfrac{3}{4}\right)^2 \div \left(\dfrac{2}{3} + \dfrac{5}{6}\right)$

27. Find the average of $\dfrac{5}{6}$, $\dfrac{4}{3}$, and $\dfrac{7}{12}$.

Simplify each complex fraction.

28. $\dfrac{\dfrac{5x}{7}}{\dfrac{20x^2}{21}}$

29. $\dfrac{5 + \dfrac{3}{7}}{2 - \dfrac{1}{2}}$

Solve.

30. $-\dfrac{3}{8}x = \dfrac{3}{4}$

31. $\dfrac{x}{5} + x = -\dfrac{24}{5}$

32. $\dfrac{2}{3} + \dfrac{x}{4} = \dfrac{5}{12} + \dfrac{x}{2}$

Evaluate each expression for the given replacement values.

33. $-5x$; $x = -\dfrac{1}{2}$

34. $x \div y$; $x = \dfrac{1}{2}$, $y = 3\dfrac{7}{8}$

Solve.

35. A carpenter cuts a piece $2\dfrac{3}{4}$ feet long from a cedar plank that is $6\dfrac{1}{2}$ feet long. How long is the remaining piece?

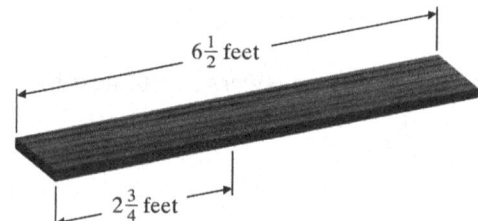

$6\frac{1}{2}$ feet

$2\frac{3}{4}$ feet

The circle graph below shows how the average consumer spends money. For example, $\frac{7}{50}$ of spending goes for food. Use this information for Exercises 36 through 38.

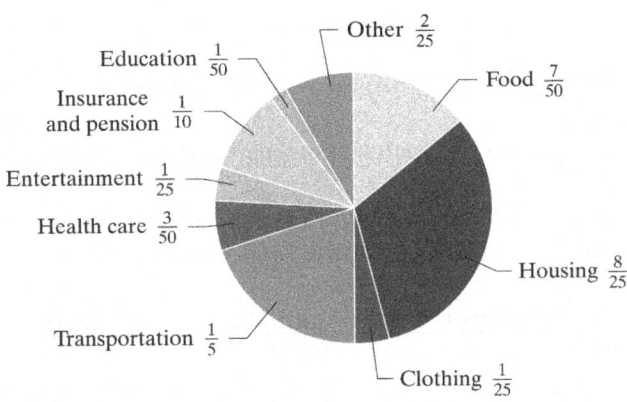

Consumer Spending

Other $\frac{2}{25}$

Education $\frac{1}{50}$

Insurance and pension $\frac{1}{10}$

Entertainment $\frac{1}{25}$

Health care $\frac{3}{50}$

Food $\frac{7}{50}$

Housing $\frac{8}{25}$

Transportation $\frac{1}{5}$

Clothing $\frac{1}{25}$

Source: U.S. Bureau of Labor Statistics; based on survey

36. What fraction of spending goes for housing and food combined?

37. What fraction of spending goes for education, transportation, and clothing?

38. Suppose your family spend $47,000 on the items in the graph. How much might we expect was spent on health care?

Find the perimeter and area of the figure.

39.

Rectangle $\frac{2}{3}$ foot

1 foot

40. During a 258-mile trip, a car used $10\frac{3}{4}$ gallons of gas. How many miles would we expect the car to travel on 1 gallon of gas?

Section 5.1 Introduction to Decimals

Objectives

    A Know the Meaning of Place Value for a Decimal Number and Write Decimals in Words
    B Write Decimals in Standard Form
    C Write Decimals as Fractions
    D Compare Decimals
    E Round Decimals to Given Place Values

Directions: Complete your Interactive Organizer by filling in the blanks and solving exercises as you complete each screen of the Interactive Assignment.

- For **WORK WITH ME** exercises, follow along and write each step needed and shown to solve, including the final answer.
- For **YOUR TURN** exercises, write the exercise generated for you in MyLab Math, then "show your work" by writing each step needed to solve, including the final answer.

Objective A: Know the Meaning of Place Value for a Decimal Number and Write Decimals in Words

Watch the objective video.

Numbers written in decimal notation are called _____ numbers, or simply _____ .

In the number 16.23, the whole number part is _____ and the decimal part is _____ .

Writing (or Reading) a Decimal in Words

  Step 1:

  Step 2:

  Step 3:

Write 16.23 in words.

Write 167.009 in words.

YOUR TURN #1:

YOUR TURN #2:

Objective B: Write Decimals in Standard Form

Watch the objective video.

VIDEO WORK WITH ME.

Write in standard form:

   Nine and eight hundredths   Forty-six ten-thousandths

YOUR TURN #1:   **YOUR TURN #2:**

Objective C: Write Decimals as Fractions

VIDEO WORK WITH ME.

Write as a fraction or mixed number.

   0.27   7.008

By reading a decimal number _____ , one can write it correctly as a fraction (or mixed number).

YOUR TURN #1:   **YOUR TURN #2:**

Objective D: Compare Decimals

Watch the objective video.

Comparing Two Positive Decimals
Compare digits in the same places from _____ to _____ . When two digits are not equal, the number with the larger digit is the _____ decimal. If necessary, insert 0s after the last digit on the _____ of the decimal point to continue comparing.

VIDEO WORK WITH ME.

Compare.

167.908   167.980

Helpful Hint #1

For any decimal, writing _____ after the last digit to the right of the decimal point does not change the value of the number.

$$7.6 =$$

Helpful Hint #2

When a whole number is written as a decimal, the decimal point is placed to the _____ of the ones digit.

$$25 =$$

Helpful Hint #3

If you have trouble comparing two negative decimals, try the following: Compare their _____ values. Then correctly compare the negative decimals, _____ the direction of the inequality symbol.

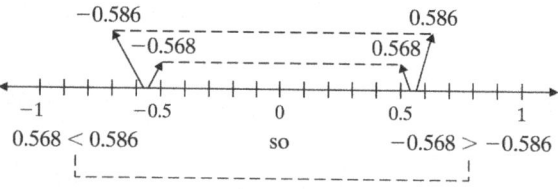

$$0.568 < 0.586 \qquad \text{so} \qquad -0.568 > -0.586$$

WORK WITH ME #1.

YOUR TURN #1:   **YOUR TURN #2:**

Objective E: Round Decimals to Given Place Values

Watch the objective video.

Rounding Decimals to a Place Value to the Right of the Decimal Point

Step 1:

Step 2:

VIDEO WORK WITH ME.

Round 0.57 to the nearest tenth.　　　　　　　Round 0.5942 to the nearest thousandth.

WORK WITH ME #1.

YOUR TURN #1:　　　　　　　　　**YOUR TURN #2:**

YOUR TURN #3:

Section 5.2 Adding and Subtracting Decimals

Objectives
A Add or Subtract Decimals
B Estimate When Adding or Subtracting Decimals
C Evaluate Expressions with Decimal Replacement Values
D Simplify Expressions Containing Decimals
E Solve Problems That Involve Adding or Subtracting Decimals

Directions: Complete your Interactive Organizer by filling in the blanks and solving exercises as you complete each screen of the Interactive Assignment.

- For **WORK WITH ME** exercises, follow along and write each step needed and shown to solve, including the final answer.
- For **YOUR TURN** exercises, write the exercise generated for you in MyLab Math, then "show your work" by writing each step needed to solve, including the final answer.

Objective A: Add or Subtract Decimals

Watch the objective video.

Adding or Subtracting Decimals
Step 1:
Step 2:
Step 3:

VIDEO WORK WITH ME.

$24.6 + 2.39 + 0.0678$

$$\begin{array}{r} 654.9 \\ -\ 56.67 \\ \hline \end{array}$$

$-1.12 - 5.2$

YOUR TURN #1:   **YOUR TURN #2:**

YOUR TURN #3:   **YOUR TURN #4:**

Objective B: Estimate When Adding or Subtracting Decimals

Watch the objective video.

VIDEO WORK WITH ME.

$$\begin{array}{r} 1000 \\ -\ 123.4 \\ \hline \end{array}$$

Check subtraction by _____ .

YOUR TURN #1:   **YOUR TURN #2:**

Objective C: Evaluate Expressions with Decimal Replacement Values

Watch the objective video.

VIDEO WORK WITH ME.

Evaluate $x - z$ for $x = 3.6$ and $z = 0.21$.

YOUR TURN #1:

Objective D: Simplify Expressions Containing Decimals

Watch the objective video.

VIDEO WORK WITH ME.

Simplify:
$30.7x + 17.6 - 23.8x - 10.7$

YOUR TURN #1:

Objective E: Solve Problems That Involve Adding or Subtracting Decimals

Watch the objective video.

***VIDEO* WORK WITH ME.**

A landscape architect is planning a border for a flower garden shaped like a triangle. The sides of the garden measure 12.4 feet, 29.34 feet, and 25.7 feet. Find the amount of border material needed.

YOUR TURN #1:   **YOUR TURN #2:**

YOUR TURN #3:

Section 5.3 Multiplying Decimals and Circumference of a Circle

Objectives
A Multiply Decimals
B Estimate When Multiplying Decimals
C Multiply Decimals by Powers of 10
D Evaluate Expressions with Decimal Replacement Values
E Find the Circumference of a Circle
F Solve Problems by Multiplying Decimals

Directions: Complete your Interactive Organizer by filling in the blanks and solving exercises as you complete each screen of the Interactive Assignment.

- For **WORK WITH ME** exercises, follow along and write each step needed and shown to solve, including the final answer.
- For **YOUR TURN** exercises, write the exercise generated for you in MyLab Math, then "show your work" by writing each step needed to solve, including the final answer.

Objective A: Multiply Decimals

Watch the objective video.

Multiplying Decimals
Step 1:
Step 2:

VIDEO **WORK WITH ME.**

6.8×4.2 $\qquad\qquad\qquad\qquad$ $(-2.3)(7.65)$

YOUR TURN #1: $\qquad\qquad\qquad\qquad$ **YOUR TURN #2:**

Objective B: Estimate When Multiplying Decimals

Watch the objective video.

VIDEO **WORK WITH ME.**

6.8×4.2

WORK WITH ME #1.

Multiply. 28.06×1.95

YOUR TURN #1:

Objective C: Multiply Decimals by Powers of 10

There are patterns that occur when we multiply a number by a power of 10 such as 10, 100, 1000, 10,000 and so on.

$23.6951 \times 100 = 2369.51$

2 zeros   Move the decimal point *2 places* to the *right.*

$23.6951 \times 100,000 = 2,369,510.$

5 zeros   Move the decimal point *5 places* to the *right* (insert a 0).

Multiplying Decimals by Power of 10 such as 10, 100, 1000, 10,000

Move the decimal point to the _____ the same number of places as there are _____ in the power of 10.

WORK WITH ME #1.
Multiply.

a. 6.5×10

b. $(147.9)(100)$

There are also patterns that occur when we multiply a number by a power of 10 that is less than 1, such as 0.1, 0.01, 0.001, 0.0001, and so on.

$569.2 \times 0.1 = 56.92$

1 decimal place   Move the decimal point *1 place* to the *left.*

$569.2 \times 0.0001 = 0.05692$

4 decimal places   Move the decimal point *4 places* to the *left* (insert one 0).

Multiplying Decimals by Power of 10 such as 0.1, 0.01, 0.001, 0.0001

Move the decimal point to the _____ the same number of places as there are _____ places in the power of 10.

112

WORK WITH ME #2.
Multiply.
 a. 8.3×0.1

 b. $(-9.83)(-0.01)$

YOUR TURN #1:   **YOUR TURN #2:**

YOUR TURN #3:

Objective D: Evaluate Expressions with Decimal Replacement Values

Watch the objective video.

***VIDEO* WORK WITH ME.**

Evaluate $xz - y$ for $x = 3$, $y = -0.2$, and $z = 5.7$

YOUR TURN #1:

Objective E: Find the Circumference of a Circle

Watch the objective video.

Circumference of a Circle

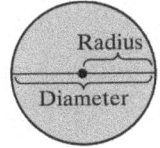 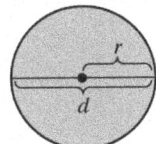

Circumference $= 2 \cdot \pi \cdot \text{radius}$ or
Circumference $= \pi \cdot \text{diameter}$
$C = 2\pi r$ or $C = \pi d$

The _____ of a circle is the distance around the circle.

***VIDEO* WORK WITH ME.**

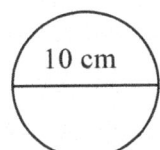

YOUR TURN #1:

Objective F: Solve Problems by Multiplying Decimals

Watch the objective video.

VIDEO WORK WITH ME.

A 1-ounce serving of cream cheese contains 6.2 grams of saturated fat. How much saturated fat is in 4 ounces of cream cheese?

YOUR TURN #1:

Section 5.4 Dividing Decimals

Objectives

    A Dividing Decimals
    B Estimate When Dividing Decimals
    C Divide Decimals by Powers of 10
    D Evaluate Expressions with Decimal Replacement Values
    E Solve Problems by Dividing Decimals

Directions: Complete your Interactive Organizer by filling in the blanks and solving exercises as you complete each screen of the Interactive Assignment.
- For **WORK WITH ME** exercises, follow along and write each step needed and shown to solve, including the final answer.
- For **YOUR TURN** exercises, write the exercise generated for you in MyLab Math, then "show your work" by writing each step needed to solve, including the final answer.

Objective A: Dividing Decimals

Dividing decimal numbers is similar to dividing _____ numbers. The only difference is that we place a decimal point in the _____ .

Dividing by a Decimal

Step 1

Step 2

Step 3

WORK WITH ME #1.

Divide.
a. $4.756 \div 0.82$

b. Divide $68.39 \div 0.6$. Round the quotient to the nearest tenth.

YOUR TURN #1:   **YOUR TURN #2:**

Objective B: Estimate When Dividing Decimals

Watch the objective video.

VIDEO WORK WITH ME.

$$5.5\overline{)36.3}$$

YOUR TURN #1:   **YOUR TURN #2:**

Objective C: Divide Decimals by Powers of 10

Watch the objective video.

> **Dividing Decimals by Powers of 10 such as 10, 100, or 1000**
>
> Move the decimal point of the dividend to the _____ the same number of places as there are _____ in the power of 10.

VIDEO WORK WITH ME.

$12.9 \div (-1000)$

YOUR TURN #1:   **YOUR TURN #2:**

Objective D: Evaluate Expressions with Decimal Replacement Values

Watch the objective video.

If a _____ statement results, the number is a solution.

If a true statement does not result, the number is _____ a solution.

VIDEO WORK WITH ME.

Determine whether $x = 12.16$ is a solution of $\frac{x}{4} = 3.04$.

YOUR TURN #1:

YOUR TURN #2:

Objective E: Solve Problems by Dividing Decimals

Watch the objective video.

VIDEO WORK WITH ME.

In the United States, an average child will wear down 730 crayons by his or her tenth birthday. Find the number of boxes of 64 crayons this is equivalent to. Round to the nearest tenth.

YOUR TURN #1:

YOUR TURN #2:

Section 5.5 Fractions, Decimals, and Order of Operations

Objectives

        A Write Fractions as Decimals

        B Compare Decimals and Fractions

        C Simplify Expressions Containing Decimals and Fractions Using Order of Operations

        D Solve Area Problems Containing Fractions and Decimals

        E Evaluate Expressions Given Decimal Replacement Values

Directions: Complete your Interactive Organizer by filling in the blanks and solving exercises as you complete each screen of the Interactive Assignment.

- For **WORK WITH ME** exercises, follow along and write each step needed and shown to solve, including the final answer.
- For **YOUR TURN** exercises, write the exercise generated for you in MyLab Math, then "show your work" by writing each step needed to solve, including the final answer.

Objective A: Write Fractions as Decimals

To write a fraction as a decimal, we interpret the fraction _____ to mean division and find the quotient.

Writing Fractions as Decimals

To write a fraction as a decimal, _____ the numerator by the denominator.

An Example

To write $\dfrac{1}{4}$ as a decimal, we divide.

Thus,

WORK WITH ME #1.

Write each number as a decimal.

a. $\dfrac{3}{4}$

b. $\dfrac{11}{12}$

YOUR TURN #1:

YOUR TURN #2:

Objective B: Compare Decimals and Fractions

Watch the objective video.

VIDEO WORK WITH ME.

Compare: $1.3 \quad \dfrac{18}{13}$

YOUR TURN #1:   **YOUR TURN #2:**

Objective C: Simplify Expressions Containing Decimals and Fractions Using Order of Operations

Watch the objective video.

Order of Operations
Step 1:
Step 2:
Step 3:
Step 4:

VIDEO WORK WITH ME.

$$\dfrac{7 + 0.74}{-6}$$

YOUR TURN #1:

Objective D: Solve Area Problems Containing Fractions and Decimals

Watch the objective video.

VIDEO WORK WITH ME.

0.62 yd

$\dfrac{2}{5}$ yd

WORK WITH ME #1.

YOUR TURN #1:

Objective E: Evaluate Expressions Given Decimal Replacement Values

Watch the objective video.

VIDEO WORK WITH ME.

Evaluate $4y - z$ for $y = 0.3$ and $z = -2.4$.

YOUR TURN #1:

120

Section 5.6 Solving Equations Containing Decimals

Objectives
   A Solve Equations Containing Decimals

Directions: Complete your Interactive Organizer by filling in the blanks and solving exercises as you complete each screen of the Interactive Assignment.
- For **WORK WITH ME** exercises, follow along and write each step needed and shown to solve, including the final answer.
- For **YOUR TURN** exercises, write the exercise generated for you in MyLab Math, then "show your work" by writing each step needed to solve, including the final answer.

Objective A: Solve Equations Containing Decimals

We continue our work with decimals by solving equations containing decimals. Let's review our steps for solving an equation.

Steps for Solving an Equation in x

Step 1

Step 2

Step 3

Step 4

Step 5

Step 6

WORK WITH ME #1.

Solve.
   a. $6x + 8.65 = 3x + 10$ b. $3(x + 2.71) = 2x$

YOUR TURN #1:   **YOUR TURN #2:**

YOUR TURN #3:

WORK WITH ME #2.

Solve.

$1.2 + 0.3x = 0.9$

When solving equations with decimals, sometimes it may be easier to first rewrite the equation so that it contains _____ only.

To do this, we can use the _____ property of equality to multiply both sides of the equation by an appropriate power of 10.

WORK WITH ME #3.

Solve: $0.5y + 2.3 = 1.65$

YOUR TURN #4:

YOUR TURN #5:

Section 5.7 Decimal Applications: Mean, Median, and Mode

Objectives
A Find the Mean of a List of Numbers
B Find the Median of a List of Numbers
C Find the Mode of a List of Numbers

Directions: Complete your Interactive Organizer by filling in the blanks and solving exercises as you complete each screen of the Interactive Assignment.
- For **WORK WITH ME** exercises, follow along and write each step needed and shown to solve, including the final answer.
- For **YOUR TURN** exercises, write the exercise generated for you in MyLab Math, then "show your work" by writing each step needed to solve, including the final answer.

Objective A: Find the Mean of a List of Numbers

Sometimes it is desirable to be able to describe a set of data, or a set of numbers, by a single "middle" number. Three such **measures of _____ tendency** are the **mean**, the **median**, and the **mode.**

Let's begin with a formula for calculating the mean (average).

The **mean (average)** of a set of numbered items is the sum of the items divided by the number of items.
mean = _____

WORK WITH ME #1.

Find the mean of the following list of numbers. If necessary, round the mean to one decimal place.
    7.6, 8.2, 8.2, 9.6, 5.7, 9.1

YOUR TURN #1:

Often in college, the calculation of a _____ **point average** (GPA) is a _____ **mean** and is calculated as shown next.

WORK WITH ME #2.

Grades for a student are shown for a particular semester. Find the grade point average. If necessary, round the grade point average to the nearest hundredth.

Grade	Credit Hours
B	3
C	3
A	4
C	4

YOUR TURN #2:

Objective B: Find the Median of a List of Numbers

Watch the objective video.

The _____ of a set of numbers in numerical order is the middle number.
       If the number of items is odd, the median is the _____ number.
       If the number of items is even, the median is the _____ of the two middle numbers.

To find the median, first place the numbers in _____ order.

VIDEO WORK WITH ME.

7.6, 8.2, 8.2, 9.6, 5.7, 9.1   mean: 8.1

YOUR TURN #1:

Objective C: Find the Mode of a List of Numbers

Watch the objective video.

The _____ of a set of numbers is the number that occurs most _____ . (It is possible for a set of numbers to have more than one mode or to have no mode.)

VIDEO WORK WITH ME.

7.6, 8.2, 8.2, 9.6, 5.7, 9.1   mean: 8.1   median: 8.2

Helpful Hint
Don't forget that it is possible for a list of numbers to have no mode or more than one mode.

Example
          The list 2, 4, 5, 6, 8, 9 has _____ mode.

          The list 6, 8, 8, 8, 11, 11, 15, 19, 19, 19, 25, 27, 27 has _____ modes.
          They are _____ and _____ .

YOUR TURN #1:

Chapter 5 Review and Practice

Study Skills
Chapter Vocabulary
Getting Ready for the Test
Review Exercises
Practice Chapter Test

Study Skills

Directions: **Watch the Study Skills video.**

Chapter Vocabulary

WORK WITH ME.

Fill in each blank with one of the words or phrases listed below:

vertically	decimal	and	right triangle	standard form	mean

median	circumference	sum	denominator	numerator	mode

1. Like fractional notation, _____ notation is used to denote a part of a whole.

2. To write fractions as decimals, divide the _____ by the _____ .

3. To add or subtract decimals, write the decimals so that the decimal points line up
_____ .

4. When writing decimals in words, write " _____ " for the decimal point.

5. When multiplying decimals, the decimal point in the product is placed so that the number of decimal places in the product is equal to the _____ of the number of decimal places in the factors.

6. The _____ of a set of numbers is the number that occurs most often.

7. The distance around a circle is called the _____ .

8. The _____ of a set of numbers in numerical order is the middle number. If there is an even number of numbers, the median is the _____ of the two middle numbers.

9. The _____ of a list of items with number values is $\dfrac{\text{sum of items}}{\text{number of items}}$.

10. When 2 million is written as 2,000,000, we say it is written in _____ .

Getting Ready for the Test.

- These exercises will help you avoid common errors while taking your chapter test.

General Directions: Read the exercise Write any notes or steps in this Interactive Organizer, along with your answer to the exercise. In the MyLab Math Interactive Assignment, click the **SHOW ANSWERS** button to check your answers. Correct any errors, or press the **PLAY** button for a video solution.

Multiple Choice. Al the exercises are Multiple Choice. Choose the correct letter.

For Exercises 1 through 4, name the place value for the given digit. The number is 1026.89704 and choices are:
  A. tens   B. hundreds   C. tenths   D. thousandths   E. ten-thousandths

1. the digit 8   2. the digit 2

3. the digit 0 to the left of the decimal point   4. the digit 0 to the right of the decimal point

For Exercises 5 through 8, let $x = 3$ and $y = 0.2$. Choose the expression that gives each answer.
  A. $x + y$ B. $x - y$ C. xy D. $x \div y$

5. answer: 0.6   6. answer: 3.2

7. answer: 2.8   8. answer: 15

For Exercises 9 through 11, choose the correct letter.

9. The expression $2(0.5)^2 + 0.3$ simplifies to
        A. 1.3   B. 0.8   C. 2.3   D. 1.28

10. One way to solve $5(x - 3.2) = 9.5$ is to first use the distributive property to multiply on the left side of the equation. Once this is done, the equivalent equation is
        A. $5x - 3.2 = 9.5$ B. $5x - 16 = 9.5$ C. $5x - 16 = 95$

11. One way to solve $0.3x + 1.8 = 2.16$ is to first multiply both sides of the equation by 100. Once this is done, the equivalent equation is
        A. $3x + 18 = 216$ B. $3x + 1.8 = 216$ C. $30x + 180 = 216$

For Exercises 12 through 14, choose the correct directions that lead to the given correct answer for the data set: 7, 9, 10, 13, 13
  A. Find the mean.   B. Find the median.   C. Find the mode.

12. answer: 10   13. answer: 13   14. answer: 10.4

Review Exercises

In the **MyLab Math, Interactive Assignment, Review Exercises** section, there are algorithmically generated "Your Turn" exercises so that you can check your knowledge of some core concepts in this chapter. Insert a few sheets of paper in your Interactive Organizer to "record and show your work" along with the final answer.

Practice Chapter Test

- These exercises will help you practice for your chapter test.

General Directions: For each exercise, "show your work" by writing each step in the solution process within your Interactive Organizer, including your final answer. In the MyLab Math Interactive Assignment, click the Show Answer button to check your answer. Correct any errors, or press the **PLAY** button for a video solution.

Write each decimal as indicated.
1. 45.092, in words

2. Three thousand and fifty-nine thousandths, in standard form

Perform each indicated operation. Round the result to the nearest thousandth if necessary.
3. $2.893 + 4.21 + 10.492$ 4. $-47.92 - 3.28$ 5. $9.83 - 30.25$

6. 10.2×4.01 7. $-0.00843 \div (-0.23)$

Round each decimal to the indicated place value.
8. 34.8923, nearest tenth   9. 0.8623, nearest thousandth

Insert <, <, or = between each pair of numbers to form a true statement.
10. 25.0909   25.9090   11. $\dfrac{4}{9}$ 0.445

Write each decimal as a fraction or a mixed number.
12. 0.345   13. -24.73

Write each fraction as a decimal. If necessary, round to the nearest thousandth.
14. $-\dfrac{13}{26}$ 15. $\dfrac{16}{17}$

Simplify.
16. $(-0.6)^2 + 1.57$ 17. $\dfrac{0.23 + 1.63}{-0.3}$ 18. $2.4x - 3.6 - 1.9x - 9.8$

Solve.

19. $0.2x + 1.3 = 0.7$

20. $2(x + 5.7) = 6x - 3.4$

Find the mean, median, and mode of each list of numbers.

21. 26, 32, 42, 43, 49

22. 8, 10, 16, 16, 14, 12, 12, 13

Find the grade point average. If necessary, round to the nearest hundredth.

23.

Grade	Credit Hours
A	3
B	3
C	3
B	4
A	1

Solve.

24. At its farthest, Pluto is 4583 million miles from the Sun. Write this number using standard notation.

25. Find the area.

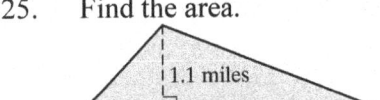

26. Find the exact circumference of the circle. Then use the spproximation 3.14 for π and approximate the circumference.

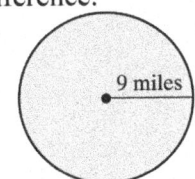

27. Vivian Thomas is going to put insecticide on her lawn to control grubworms. The lawn is a rectangle that measures 123.8 feet by 80 feet. The amount of insecticide required is 0.02 ounce per square foot.
  a. Find the area of her lawn.
  b. Find how much insecticide Vivan needs to purchase.

28. Find the total distance from Bayette to Center City.

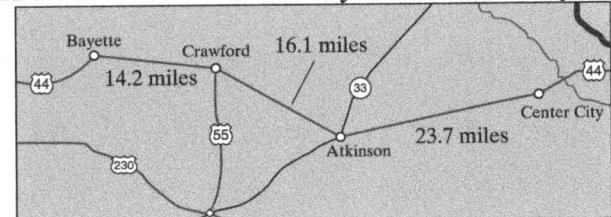

Section 6.1 Ratio and Proportion

Objectives

      A Write Ratios as Fractions

      B Solve Proportions

      C Solve Problems Modeled by Proportions

Directions: Complete your Interactive Organizer by filling in the blanks and solving exercises as you complete each screen of the Interactive Assignment.

- For **WORK WITH ME** exercises, follow along and write each step needed and shown to solve, including the final answer.
- For **YOUR TURN** exercises, write the exercise generated for you in MyLab Math, then "show your work" by writing each step needed to solve, including the final answer.

Objective A: Write Ratios as Fractions

Watch the objective video.

1 to 2 $\dfrac{1}{2}$ 1:2

Ratio

The ratio of a number *a* to a number *b* is their _____. Ways of writing ratios are

_____ , _____ , and _____

VIDEO WORK WITH ME.

10 inches to 12 inches   4 nickels to 2 dollars

YOUR TURN #1:   **YOUR TURN #2:**

YOUR TURN #3:

Objective B: Solve Proportions

Watch the objective video.

Cross Products
If $\dfrac{a}{b} = \dfrac{c}{d}$, then $ad =$ _____ .

VIDEO WORK WITH ME.

$$\dfrac{x}{10} = \dfrac{5}{9} \qquad\qquad\qquad \dfrac{x+1}{2x+3} = \dfrac{2}{3}$$

A _____ is a statement that two ratios are equal.

Make sure proposed solution does not make denominator _____ . If so, _____ solution.

YOUR TURN #1:   **YOUR TURN #2:**

Objective C: Solve Problems Modeled by Proportions

Watch the objective video.

VIDEO WORK WITH ME.

There are 110 calories per 28.8 grams of Frosted Flakes cereal. Find how many calories are in 43.2 grams of this cereal.

YOUR TURN #1:

Section 6.2 Percents, Decimals, and Fractions

Objectives
A Understand Percent
B Write Percents as Decimals or Fractions
C Write Decimals or Fractions as Percents
D Solve Applications with Percents, Decimals, and Fractions

Directions: Complete your Interactive Organizer by filling in the blanks and solving exercises as you complete each screen of the Interactive Assignment.

- For **WORK WITH ME** exercises, follow along and write each step needed and shown to solve, including the final answer.
- For **YOUR TURN** exercises, write the exercise generated for you in MyLab Math, then "show your work" by writing each step needed to solve, including the final answer.

Objective A: Understand Percent

Watch the objective video.

_____ means "per 100".

VIDEO WORK WITH ME.

In a survey of 100 college students, 96 use the Internet. What percent use the internet?

YOUR TURN #1:   **YOUR TURN #2:**

Objective B: Write Percents as Decimals or Fractions

Since percent means "per hundred," we have that

$$1\% = \frac{1}{100} =$$

Write 87% as a fraction.

Write 87% as a decimal.

Show 87% as a fraction and a decimal.

Let's first write percents as decimals.

Writing a Percent as a Decimal

Replace the percent symbol with its decimal equivalent, _____ ; then multiply.

43% =

Helpful Hint
Writing a Percent as a Decimal

If it helps, think of writing a percent as a decimal by

Percent → | Remove the _____ symbol and move the decimal point 2 places to the _____ . | → Decimal

YOUR TURN #1:   **YOUR TURN #2:**

YOUR TURN #3:

Now let's first write percents as fractions.

Writing a Percent as a Fraction

Replace the percent symbol with its fraction equivalent _____ ; then multiply. Don't forget to _____ the fraction if possible.

43% =

WORK WITH ME #1.

Write each percent as a fraction or mixed number in simplest form.

    a. 8%   b. 175%   c. $10\frac{1}{3}\%$

YOUR TURN #4:   **YOUR TURN #5:**

YOUR TURN #6:

Objective C: Write Decimals or Fractions as Percents

Now we write decimals or fractions as percents.
To do so, we use the fact that
$$1 = \underline{\hspace{1.5cm}}$$

Write 0.38 as a percent.

Write $\dfrac{1}{5}$ as a percent.

First, let's practice writing decimals as percents.

Writing a Decimal as a Percent

Multiply by 1 in the form of _____ .

  0.27 =

Helpful Hint

Writing a Decimal as a Percent

If it helps, think of writing a decimal as a percent by reversing the steps in the *Helpful Hint for Writing a Percent as a Decimal*

Percent → | Move the _____ point
2 places to the _____ and
attach a % symbol. | → Decimal

YOUR TURN #1:   **YOUR TURN #2:**

YOUR TURN #3:

Now let's write fractions as percents.

Writing a Fraction as a Percent

Multiply by 1 in the form of _____ .

$$\frac{1}{8} =$$

Helpful Hint

Don't forget that
$$100\% =$$
Recall that when we multiply a number by _____ , we are not changing the value of that number. This means that when we multiply a number by _____ , we are not changing its value but rather writing the number as an equivalent percent.

WORK WITH ME #1.

Write each fraction as a percent.

a. $\dfrac{4}{5}$ b. $\dfrac{1}{3}$

YOUR TURN #4:   **YOUR TURN #5:**

Objective D: Solve Applications with Percents, Decimals, and Fractions

Watch the objective video.

Summary of Converting Percents, Decimals, and Fractions

- *To write a percent as a decimal*, replace the % symbol with its decimal equivalent, 0.01; then multiply.

- *To write a percent as a fraction*, replace the % symbol with its fraction equivalent $\frac{1}{100}$; then multiply.

- *To write a decimal or fraction as a percent*, multiply by 100%.

VIDEO WORK WITH ME.

People take aspirin for a variety of reasons. The most common use of aspirin is to prevent heart disease, accounting for 38% of all aspirin use.

WORK WITH ME #1.

Write each fraction as a percent.

a. A family decides to spend no more than 25% of its monthly income on rent. Write 25% as a decimal.

b. An advertisement for a stereo system reads "$\frac{1}{3}$ off." What percent off is this?

YOUR TURN #1:

YOUR TURN #2:

Section 6.3 Solving Percent Problems with Equations

Objectives
A Write Percent Problems as Equations
B Solve Percent Problems

Directions: Complete your Interactive Organizer by filling in the blanks and solving exercises as you complete each screen of the Interactive Assignment.
- For **WORK WITH ME** exercises, follow along and write each step needed and shown to solve, including the final answer.
- For **YOUR TURN** exercises, write the exercise generated for you in MyLab Math, then "show your work" by writing each step needed to solve, including the final answer.

Objective A: Write Percent Problems as Equations

Watch the objective video.

"**of**" means _____   "**is**" means _____

"**what**" (or some equivalent) means the _____ number.

VIDEO **WORK WITH ME.**

18% of 81 is what number?   What percent of 80 is 3.8?

VIDEO **WORK WITH ME.**

1.2 is 12% of what number?

YOUR TURN #1:   **YOUR TURN #2:**

YOUR TURN #3:

Objective B: Solve Percent Problems

Notice that each percent problem contains _____ numbers (in our examples, two are known and one is unknown).

Each of these numbers is given a special name.

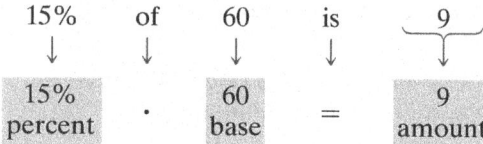

We call this equation the _____ **equation**.

Percent Equation
          percent · base = amount

WORK WITH ME #1.

Solve.
   10% of 35 is what number?

WORK WITH ME #2.

WORK WITH ME #3.

Solve.
   2.58 is what percent of 50?

WORK WITH ME #4.

Solve.
   12% of what number is 0.6?

Helpful Hint
If your unknown in the percent equation is a _____, don't forget to convert your answer to a _____ .

YOUR TURN #1:

YOUR TURN #2:

YOUR TURN #3:

Section 6.4 Solving Percent Problems with Proportions

Objectives
A Write Percent Problems as Proportions
B Solve Percent Problems

Directions: Complete your Interactive Organizer by filling in the blanks and solving exercises as you complete each screen of the Interactive Assignment.

- For **WORK WITH ME** exercises, follow along and write each step needed and shown to solve, including the final answer.
- For **YOUR TURN** exercises, write the exercise generated for you in MyLab Math, then "show your work" by writing each step needed to solve, including the final answer.

Objective A: Write Percent Problems as Proportions

Watch the objective video.

Percent Proportion
$$\frac{\rule{2cm}{0.4pt}}{\text{base}} = \frac{\text{percent}}{100}$$
or
$$\rule{1.5cm}{0.4pt} = \rule{1.5cm}{0.4pt}$$

The _____, *p*, can be identified by the looking for the % symbol.

The _____, *b*, usually follows the word _____ .

The _____, *a,* is the part compared to the whole.

VIDEO WORK WITH ME.

98% of 45 is what number?   What percent of 400 is 70?

VIDEO WORK WITH ME.

7.8 is 78% of what number?

WORK WITH ME #1.

YOUR TURN #1:   **YOUR TURN #2:**

YOUR TURN #3:

Objective B: Solve Percent Problems

The proportions we have written contain three values: the _____, the _____, and the
_____ . If any _____ of these values are known, we can find the third (the unknown
value). To do this, we write a percent proportion and _____ for the unknown value.

WORK WITH ME #1.   **WORK WITH ME #2.**

Solve.   Solve.
   7.8 is 78% of what number?   What percent of 6 is 2.7?

Helpful Hint
Recall from our percent proportion that this number already is a _____ . Just keep the number
as is and _____ a % symbol.

YOUR TURN #1:   **YOUR TURN #2:**

YOUR TURN #3:

Section 6.5 Applications of Percent

Objectives
A Solve Applications Involving Percent
B Find Percent Increase and Percent Decrease

Directions: Complete your Interactive Organizer by filling in the blanks and solving exercises as you complete each screen of the Interactive Assignment.

For **WORK WITH ME** exercises, follow along and write each step needed and shown to solve, including the final answer.

For **YOUR TURN** exercises, write the exercise generated for you in MyLab Math, then "show your work" by writing each step needed to solve, including the final answer.

Objective A: Solve Applications Involving Percent

Watch the objective video.

VIDEO WORK WITH ME.

A family paid $26,250 as a down payment for a home. If this represents 15% of the price of the home, find the price of the home.

WORK WITH ME #1.

The freshman class of 775 students is 31% of all students at Euclid University. How many students go to Euclid University?

YOUR TURN #1:

141

Objective B: Find Percent Increase and Percent Decrease

Watch the objective video.

Percent Increase   percent of increase = Then write the quotient as a percent.

Percent of Decrease   percent of decrease = Then write the quotient as a percent.

VIDEO **WORK WITH ME.**

There are 150 calories in a cup of whole milk and only 84 in a cup of skim milk. In switching to skim milk, find the percent decrease in number of calories per cup.

VIDEO **WORK WITH ME.**

In 1940, the average size of a privately owned farm in the United States was 174 acres. In a recent year, the average size of a privately owned farm in the United States had increased to 420 acres. What is this percent increase?

YOUR TURN #1:   **YOUR TURN #2:**

Section 6.6 Percent and Problem Solving: Sales Tax, Commission, and Discount

Objectives
A Calculate Sales Tax and Total Price
B Calculate Commissions
C Calculate Discount and Sale Price

Directions: Complete your Interactive Organizer by filling in the blanks and solving exercises as you complete each screen of the Interactive Assignment.

For **WORK WITH ME** exercises, follow along and write each step needed and shown to solve, including the final answer.

For **YOUR TURN** exercises, write the exercise generated for you in MyLab Math, then "show your work" by writing each step needed to solve, including the final answer.

Objective A: Calculate Sales Tax and Total Price

Watch the objective video.

Sales Tax and Total Price
sales tax =
total price =

VIDEO **WORK WITH ME.**

The sales tax on the purchase of a futon is $24.25. If the tax rate is 5%, find the purchase price of the futon.

YOUR TURN #1:   **YOUR TURN #2:**

Objective B: Calculate Commissions

Watch the objective video.

Commission
commission =

VIDEO WORK WITH ME.

A salesperson earned a commission of $1380.40 for selling $9860 worth of paper products. Find the commission rate.

YOUR TURN #1:

Objective C: Calculate Discount and Sale Price

Watch the objective video.

Discount and Sale Price
amount of discount =
sale price =

VIDEO WORK WITH ME.

A $300 fax machine is on sale for 15% off. Find the amount of discount and the sale price.

YOUR TURN #1:

Section 6.7 Percent and Problem Solving: Interest

Objectives
     A Calculate Simple Interest
     B Calculate Compound Interest

Directions: Complete your Interactive Organizer by filling in the blanks and solving exercises as you complete each screen of the Interactive Assignment.

- For **WORK WITH ME** exercises, follow along and write each step needed and shown to solve, including the final answer.
- For **YOUR TURN** exercises, write the exercise generated for you in MyLab Math, then "show your work" by writing each step needed to solve, including the final answer.

Objective A: Calculate Simple Interest

Watch the objective video.

Simple Interest

     Simple Interest =

$$I = P \cdot R \cdot T$$

where the rate is understood to be per year and time is in years.

Finding the Total Amount of a Loan or Investment

     total amount (paid or received) =

VIDEO WORK WITH ME.

A company borrows $162,500 for 5 years at a simple interest rate of $12.5%. Find the interest paid on the loan and the total amount paid back.

YOUR TURN #1:

145

Objective B: Calculate Compound Interest

Watch the objective video.

Compound Interest Formula

The total amount A in an account is given by

$$A = P\left(1 + \frac{r}{n}\right)^{n \cdot t}$$

where P is the _____ , r is the interest rate written as a decimal, t is the length of
_____ in years, and n is the number of times compounded per year.

***VIDEO* WORK WITH ME.**

Find the total amount in this account. $6150 is compounded semiannually at a rate of 14%
for 15 years.

YOUR TURN #1:   **YOUR TURN #2:**

Chapter 6 Review and Practice

> Study Skills
> Chapter Vocabulary
> Getting Ready for the Test
> Review Exercises
> Practice Chapter Test

Study Skills

Directions: **Watch the Study Skills video.**

Chapter Vocabulary

WORK WITH ME.

Fill in each blank with one of the words or phrases listed below:

| percent | sales tax | is | amount of discount | percent of decrease | total price | ratio |
| proportion | of | percent of increase | sale price | commission |

1. In a mathematical statement, _____ usually means "multiplication."

2. In a mathematical statement, _____ means "equal."

3. _____ mean "per hundred."

4. _____ $= \dfrac{\text{amount of decrease}}{\text{original amount}}$.

5. _____ $= \dfrac{\text{amount of increase}}{\text{original amount}}$.

6. _____ = tax rate · purchase price

7. _____ = purchase price + sales tax

8. _____ = commission rate · sales

9. _____ = discount rate · original price

10. _____ = original price − amount of discount

11. A(n) _____ is a mathematical statement that two ratios are equal.

12. A(n) _____ is the quotient of two numbers or two quantities.

147

Getting Ready for the Test.

- These exercises will help you avoid common errors while taking your chapter test.

General Directions: Read the exercise Write any notes or steps in this Interactive Organizer, along with your answer to the exercise. In the MyLab Math Interactive Assignment, click the **SHOW ANSWERS** button to check your answers. Correct any errors, or press the **PLAY** button for a video solution.

Multiple Choice. All of the exercises are Multiple Choice. Choose the correct letter for each exercise.

1. The phrase "6 is to 7 as 18 is to 21" translates to

   A. $\dfrac{6}{7} = \dfrac{18}{21}$ B. $\dfrac{6}{18} = \dfrac{21}{7}$ C. $\dfrac{6}{21} = \dfrac{7}{18}$

2. Use cross products to determine which proportion is *not* equivalent to the proportion $\dfrac{3}{x} = \dfrac{5}{11}$.

   A. $\dfrac{3}{11} = \dfrac{x}{5}$ B. $\dfrac{x}{3} = \dfrac{11}{5}$ C. $\dfrac{3}{5} = \dfrac{x}{11}$ D. $\dfrac{11}{x} = \dfrac{5}{3}$

3. Since "percent" means "per hundred," choose the number that does *not* equal 12%.

   A. $\dfrac{12}{100}$ B. 0.12   C. $\dfrac{3}{25}$ D. 1.2

4. Choose the number that does *not* equal 100%.

   A. 10   B. 1   C. 1.00   D. $\dfrac{100}{100}$

5. Choose the number that does *not* equal 50%.

   A. 0.5   B. 5   C. $\dfrac{1}{2}$ D. $\dfrac{50}{100}$

6. Choose the number that does *not* equal 80%.

   A. 0.80   B. 0.8   C. 8   D. $\dfrac{4}{5}$

Use the information below for Exercises 7 through 10.

- 100% of a number is that original number.

- 50% of a number is half that number.

- 25% of a number is $\frac{1}{4}$ of that number.

- 10% of a number is $\frac{1}{10}$ of that number.

For Exercises 7 through 10, choose the letter that correctly fills in each blank.
  A. 100%   B. 50%   C. 25%   D. 10%

7. _____ of 70 is 35.

8. _____ of 88 is 8.8.

9. _____ of 47 is 47.

10. _____ of 28 is 7.

For Exercise 11, choose the correct letter.

11. Your bill at a restaurant is $24.86. If you want to leave a 20% tip, which tip amount is closest to 20%?
    A. $2.48   B. $20   C. $5   D. $10

For Exercises 12 and 13, an amount of $150 is to be discounted by 10%.

12. Find the amount of discount.
    A. $15   B. $45   C. $30   D. $10

13. Find the new discounted price (original price – discount).
    A. $120   B. $10   C. $140   D. $135

For Exercise 14, choose the correct letter.

14. If the original price of a pair of shoes is $40 and the shoe price is to be discounted by 25% at the register, choose the closest amount for your shoes before tax.
    A. $10   B. $20   C. $30   D. $40

Review Exercises

In the **MyLab Math, Interactive Assignment, Review Exercises** section, there are algorithmically generated "Your Turn" exercises so that you can check your knowledge of some core concepts in this chapter. Insert a few sheets of paper in your Interactive Organizer to "record and show your work" along with the final answer.

Practice Chapter Test

- These exercises will help you practice for your chapter test.

General Directions: For each exercise, "show your work" by writing each step in the solution process within your Interactive Organizer, including your final answer. In the MyLab Math Interactive Assignment, click the Show Answer button to check your answer. Correct any errors, or press the **PLAY** button for a video solution.

Write each percent as a decimal.

1. 85%   2. 500%   3. 0.6%

Write each decimal as a percent.

4. 0.056   5. 6.1   6. 0.35

Write each percent as a fraction or a mixed number in simplest form.

7. 120%   8. 38.5%   9. 0.2%

Write each fraction or mixed number as a percent.

10. $\dfrac{11}{20}$ 11. $\dfrac{3}{8}$ 12. $1\dfrac{3}{4}$

13.   In the past decade, Americans have increased their consumption of bottled water by $\dfrac{1}{4}$. Write $\dfrac{1}{4}$ as a percent. (*Source*: Bottled Water Organization)

14.   As of 2016, 45% of all working age households have no retirement plan in place. Write 45% as a fraction. (*Source*: National Institute on Retirement Security)

Solve.

15. What number is 42% of 80?

16. 0.6% of what number is 7.5?

17. 567 is what percent of 756?

Solve. If necessary, round percents to the nearest tenth, dollar amounts to the nearest cent, and all other numbers to the nearest whole.

18. An alloy is 12% copper. How much copper is contained in 320 pounds of this alloy?

19. A farmer in Nebraska estimates that 20% of his potential crop, or $11,350, has been lost to a hard freeze. Find the total value of his potential crop.

20. If the local sales tax rate is 8.25%, find the total amount charged for a stereo system priced at $354.

21. A town's population increased from 25,200 to 26,400. Find the percent of increase.

22. A $120 framed picture is on sale for 15% off. Find the discount and the sale price.

23. Randy Nguyen is paid a commission rate of 4% on all sales. Find Randy's commission if his sales were $9875.

24. A sales tax of $13.77 is added to an item's price of $152.99. Find the sales tax rate. Round to the nearest whole percent.

25. Find the simple interest earned on $2000 saved for $3\frac{1}{2}$ years at an interest rate of 9.25%.

26. $1365 is compounded annually at 8%. Find the total amount in the account after 5 years.

27. A couple borrowed $400 from a bank at 13.5% simple interest for 6 months for car repairs. Find the total amount due the bank at the end of the 6-month period.

28. In a recent 5-year period, the number of category of crimes reported in New York City decreased from 51,209 to 50,008. Find the percent of decrease. (*Source*: New York State Division of Criminal Justice Services)

29. Write the ratio $75 to $10 as a fraction in simplest form.

30. Solve: $\dfrac{5}{y+1} = \dfrac{4}{y+2}$

31. In a sample of 85 fluorescent bulbs, 3 were found to be defective. At this rate, how many defective bulbs should be found in 510 bulbs?

151

Section 7.1 Pictographs, Bar Graphs, Histograms, Line Graphs, and Introduction to Statistics

Objectives
  A Read Pictographs
  B Read and Construct Bar Graphs
  C Read and Construct Histograms for Frequency Distribution Graphs
  D Read Line Graphs
  E Calculate Range, Mean, Median, and Mode from a Frequency Distribution Table or Graph

Directions: Complete your Interactive Organizer by filling in the blanks and solving exercises as you complete each screen of the Interactive Assignment.
  - For **WORK WITH ME** exercises, follow along and write each step needed and shown to solve, including the final answer.
  - For **YOUR TURN** exercises, write the exercise generated for you in MyLab Math, then "show your work" by writing each step needed to solve, including the final answer.

Objective A: Read Pictographs

Watch the objective video.

A _____ uses pictures or symbols to show data.

VIDEO WORK WITH ME.

What is the key?

Approximate the number of wildfires in 2013.

Which year had the most wildfires?

What was the amount of decrease of wildfires from 2012 to 2013?

What was the average annual number of wildfires 2013 to 2015?

YOUR TURN #1:   **YOUR TURN #2:**

YOUR TURN #3:

Objective B: Read and Construct Bar Graphs

Watch the objective video.

_____ graphs have horizontal or vertical bars to show data.

VIDEO WORK WITH ME.

Draw the bar graph.

YOUR TURN #1:   **YOUR TURN #2:**

YOUR TURN #3:

Objective C: Read and Construct Histograms for Frequency Distribution Graphs

Watch the objective video.

A _____ is a special bar graph.

A histogram is different from a bar graph in that the width of each bar stands for a range of numbers, called a _____ **interval**.

The height of each bar is how many times a number occurs in the class and is called the **class**
_____ .

VIDEO WORK WITH ME.

How many adults drive fewer than 150 miles per week?

How many more adults drive 250-299 miles per week than 200-249 miles per week?

VIDEO WORK WITH ME.

```
78   84  91  93   97
97   95  85  95   96
101  89  92  89  100
```

Class Intervals	Tally	Class Frequency

YOUR TURN #1:   **YOUR TURN #2:**

YOUR TURN #3:   **YOUR TURN #4:**

Objective D: Read Line Graphs

Watch the objective video.

VIDEO **WORK WITH ME.**

What is the average number of goals per game in 2015?

During what year shown was the average number of goals per game the highest?

From 2013 to 2015, did the average number of goals per game increase or decrease?

During what year or years shown, were the average number of goals shown less than 8?

YOUR TURN #1:   **YOUR TURN #2:**

YOUR TURN #3:

Objective E: Calculate Range, Mean, Median, and Mode from a Frequency Distribution Table or Graph

Recall that mean, median, and mode are three measures of _____.

In this objective, we review mean, median, and mode, and we introduce _____ —one way to measure dispersion.

In statistics, _____ is a way to describe the degree to which the data values are

_____ .

Range

The range of a data set is the difference between the largest value and the smallest data value.

Range = _____ data value – _____ data value

The following pulse rates (for 1 minute) were recorded for a group of 15 students. Find the range.

78, 80, 66, 68, 71, 64, 82, 71, 70, 65, 70, 65, 70, 75, 77, 86, 72

range =

The range of this data set is _____ .

YOUR TURN #1:

Before we review mean, median, and mode, let's introduce a formula for finding the _____ of the median.

Recall that the _____ of a set of numbers in numerical order is the _____ number.

If the number of items is odd, the median is the _____ number.

If the number of items is even, the middle is the _____ (average) of the _____ middle numbers.

For a long list of data items, this formula gives us the _____ of the median.

Position of the Median

For n data items in order from _____ to _____ , the median is the item in the $\frac{n+1}{2}$ position.

Note:

If n is an even number, then the *position* formula $\frac{n+1}{2}$, will not be a _____ number.

In this case, simply find the average of the _____ data items whose positions are closest to but before and after $\frac{n+1}{2}$.

Helpful Hint

The formula above, $\frac{n+1}{2}$, does not give the _____ of the median, just the _____ of the median.

WORK WITH ME #1.

Use the frequency distribution table to find the **a.** mean, **b.** median, and **c.** mode. If needed, round the mean to 1 decimal place.

Data Item	Frequency
5	1
6	1
7	2
8	5
9	6
10	2

YOUR TURN #2:

Section 7.2 Circle Graphs

Objectives
A Read Circle Graphs
B Draw Circle Graphs

Directions: Complete your Interactive Organizer by filling in the blanks and solving exercises as you complete each screen of the Interactive Assignment.

- For **WORK WITH ME** exercises, follow along and write each step needed and shown to solve, including the final answer.
- For **YOUR TURN** exercises, write the exercise generated for you in MyLab Math, then "show your work" by writing each step needed to solve, including the final answer.

Objective A: Read Circle Graphs

Watch the objective video.

***VIDEO* WORK WITH ME.**

Where do most of these college students live?

Find the ratio of students living in campus housing to total students.

***VIDEO* WORK WITH ME.**

What percent of books are classified as some type of fiction?

What is the second largest category of books?

If the library has 125,600 books, how many of these are nonfiction?

WORK WITH ME #1.

YOUR TURN #1:   **YOUR TURN #2:**

YOUR TURN #3:   **YOUR TURN #4:**

Objective B: Draw Circle Graphs

Watch the objective video.

Type of Apple	Percent	Degrees in Sector
Red Delicious	37%	
Golden Delicious	13%	
Fuji	14%	
Gala	15%	
Granny Smith	12%	
Other varieties	6%	
Braeburn	3%	

YOUR TURN #1:

Section 7.3 Square Roots and the Pythagorean Theorem

Objectives
A Find the Square Root of a Number
B Approximate Square Roots
C Use the Pythagorean Theorem

Directions: Complete your Interactive Organizer by filling in the blanks and solving exercises as you complete each screen of the Interactive Assignment.

- For **WORK WITH ME** exercises, follow along and write each step needed and shown to solve, including the final answer.
- For **YOUR TURN** exercises, write the exercise generated for you in MyLab Math, then "show your work" by writing each step needed to solve, including the final answer.

Objective A: Find the Square Root of a Number

Watch the objective video.

The square of 7 is _____ .   The square of –7 is _____ .

The reverse process of squaring is finding a _____ .

A square root of 49 is _____ .

A square root of 49 is also _____ .

We use the notation $\sqrt{}$, called the _____ sign, to indicate the positive square root of a nonnegative number.

Square Root of a Number
The square root, $\sqrt{}$, of a positive number a, is the _____ number b whose square is a. In symbols, $\sqrt{a} = b,$ if $b^2 = a$ Also, $\sqrt{0} =$

VIDEO **WORK WITH ME.**

$\sqrt{4}$ $\sqrt{121}$ $\sqrt{\dfrac{1}{81}}$

YOUR TURN #1:   **YOUR TURN #2:**

YOUR TURN #3:

Objective B: Approximate Square Roots

Watch the objective video.

VIDEO WORK WITH ME.

$\sqrt{38}$ is between what two whole numbers? $\sqrt{15}$ is between what two whole numbers?

WORK WITH ME #1.

YOUR TURN #1:   YOUR TURN #2:

Objective C: Use the Pythagorean Theorem

One important application of square roots has to do with _____ triangles.

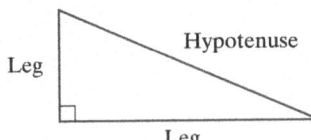

The _____ of a right triangle is the side opposite to the right angle.

A right triangle has _____ legs. Each leg is a side other than the _____ .

The right angle (_____ angle) in the triangle is indicated by the small square drawn in that angle.

Pythagorean Theorem
   If a and b are the lengths of the legs of a right triangle and c is the length of the hypotenuse, then

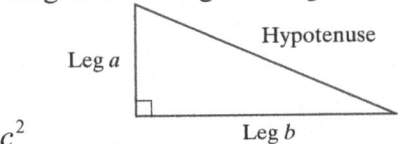

$a^2 + b^2 = c^2$

In other words, $(\text{leg})^2 + (\text{other leg})^2 = (\text{hypotenuse})^2$

161

WORK WITH ME #1.

Find the unknown length in the right triangle. If necessary, approximate the length to the nearest thousandth.

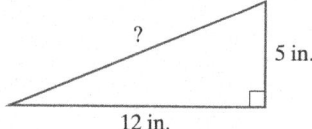

WORK WITH ME #2.

YOUR TURN #1:   **YOUR TURN #2:**

Section 7.4 Congruent and Similar Triangles

Objectives
>   A Decide Whether Two Triangles Are Congruent
>   B Find the Ratio of Corresponding Sides in Similar Triangles
>   C Find Unknown Lengths of Sides in Similar Triangles

Directions: Complete your Interactive Organizer by filling in the blanks and solving exercises as you complete each screen of the Interactive Assignment.
- For **WORK WITH ME** exercises, follow along and write each step needed and shown to solve, including the final answer.
- For **YOUR TURN** exercises, write the exercise generated for you in MyLab Math, then "show your work" by writing each step needed to solve, including the final answer.

Objective A: Decide Whether Two Triangles Are Congruent

Watch the objective video.

Congruent triangles have the same _____ and same _____ .

Congruent triangles: Corresponding *angles* have the same _____ .

                          Corresponding *sides* have the same _____ .

Congruent Triangles
Angle-Side-Angle (ASA)

If the measures of two angles of a triangle equal the measures of two angles of another triangle, and the lengths of the sides between each pair of angles are equal, the triangles are _____ .

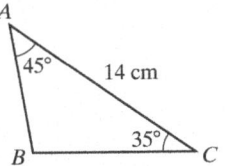
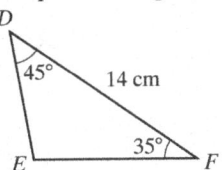

Side-Side-Side (SSS)
If the lengths of the _____ sides of a triangle equal the lengths of the corresponding sides of another triangle, the triangles are congruent.

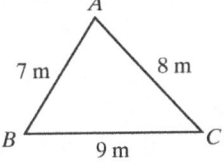
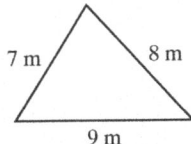

163

Side-Angle-Side (SAS)
If the lengths of two sides of a triangle equal the lengths of corresponding sides of another triangle, and the measures of the _____ between each pair of sides are equal, the triangles are congruent.

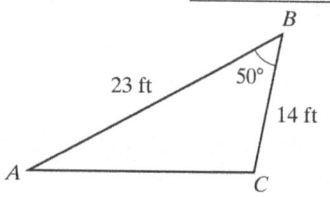 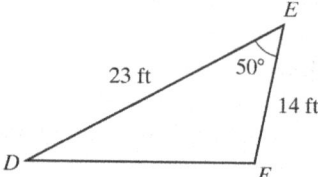

VIDEO WORK WITH ME.

Determine whether the triangles are congruent.

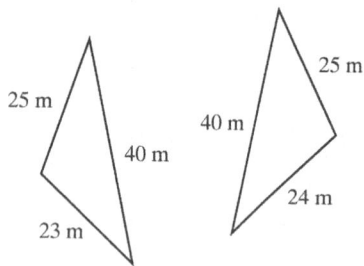

YOUR TURN #1:   **YOUR TURN #2:**

YOUR TURN #3:

Objective B: Find the Ratio of Corresponding Sides in Similar Triangles

Watch the objective video.

Similar triangles have the same _____ but not necessarily the same _____ .

VIDEO WORK WITH ME.

Find the ratio of the corresponding sides of the given similar triangles.

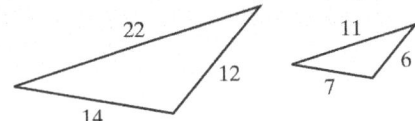

Similar triangles: Corresponding *angles* have the same _____ .

Corresponding *sides* are in _____ .

YOUR TURN #1:

Objective C: Find Unknown Lengths of Sides in Similar Triangles

Because the _____ of lengths of corresponding sides are equal, we can use _____ to find unknown lengths in similar triangles.

WORK WITH ME #1.

Write the proportion of the similar triangles.

WORK WITH ME #2.

Let's continue and find *y*. Many proportions are correct.
    We will use the proportion,

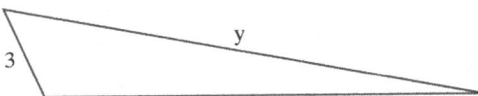

$$\frac{2}{3} = \frac{10}{y}$$

WORK WITH ME #3.

Solve.

If a 30-foot tree casts an 18-foot shadow, find the length of the shadow cast by a 24-foot tree.

YOUR TURN #1:   **YOUR TURN #2:**

Section 7.5 Counting and Introduction to Probability

Objectives
A Use a Tree Diagram to Count Outcomes
B Find the Probability of an Event

Directions: Complete your Interactive Organizer by filling in the blanks and solving exercises as you complete each screen of the Interactive Assignment.
- For **WORK WITH ME** exercises, follow along and write each step needed and shown to solve, including the final answer.
- For **YOUR TURN** exercises, write the exercise generated for you in MyLab Math, then "show your work" by writing each step needed to solve, including the final answer.

Objective A: Use a Tree Diagram to Count Outcomes

Watch the objective video.

Each chance happening is called an _____ .

The possible results on an experiment are called _____ .

VIDEO WORK WITH ME.

Choosing a letter in the word MATH, then a number (1, 2, or 3).

YOUR TURN #1:

Objective B: Find the Probability of an Event

Watch the objective video.

_____ is the measure of the chance or likelihood of an event occurring.

The Probability of an Event
$$\text{probability of an event} = \frac{\text{number of ways that the event can occur}}{\text{number of possible outcomes}}$$

VIDEO **WORK WITH ME.**

If a single die is tossed once, find the probability of a 5:

VIDEO **WORK WITH ME.**

If a single die is tossed once, find the probability of a 7:

VIDEO **WORK WITH ME.**

If a single die is tossed once, find the probability of a 1, 2, 3, 4, 5, or 6:

The probability of something impossible to happen is _____ .

The probability of something certain to happen is _____ .

VIDEO **WORK WITH ME.**

If a single die is tossed once, find the probability of an even number:

VIDEO **WORK WITH ME.**

If a spinner is spun once, what is the probability of the result of a 2?

VIDEO **WORK WITH ME.**

If a spinner is spun once, what is the probability of the result of a 4?

VIDEO **WORK WITH ME.**

If a spinner is spun once, what is the probability of the result of a 1, 2, or 3?

WORK WITH ME #1.

YOUR TURN #1:   **YOUR TURN #2:**

YOUR TURN #3:   **YOUR TURN #4:**

Chapter 7 Review and Practice

> Study Skills
> Chapter Vocabulary
> Getting Ready for the Test
> Review Exercises
> Practice Chapter Test

Study Skills

Directions: **Watch the Study Skills video.**

Chapter Vocabulary

WORK WITH ME.

Fill in each blank with one of the words or phrases listed below:

outcomes	class interval	experiment	probability	Pythagorean	leg

pictograph	bar	class frequency	hypotenuse	similar

histogram	circle	tree diagram	right	congruent

1. _____ triangles have the same shape and the same size.

2. _____ triangles have exactly the same shape but not necessarily the same size.

3. A right triangle has three sides. The longest side opposite the right angle is called the _____ , and the other two sides are each called a _____ .

4. A triangle with one right angle is called a(n) _____ triangle.

5. In the right triangle [triangle figure with sides a, b, c] , $a^2 + b^2 = c^2$ is called the _____ theorem.

6. A(n) _____ graph presents data using vertical or horizontal bars.

7. The possible results of an experiment are the _____ .

8. A(n) _____ is a graph in which pictures or symbols are used to visually present data.

9. A(n) _____ is one way to picture and count outcomes.

10. A(n) _____ is an activity being considered, such as tossing a coin or rolling a die.

11. In a(n) _____ graph, each section (shaped like a piece of pie) shows a category and the relative size of the category.

12. The _____ of an event is $\dfrac{\text{number of ways that the event can occur}}{\text{number of possible outcomes}}$.

13. A(n) _____ is a special bar graph in which the width of each bar represents a(n) _____ and the height of each bar represents the _____ .

Getting Ready for the Test.

- These exercises will help you avoid common errors while taking your chapter test.

General Directions: Read the exercise Write any notes or steps in this Interactive Organizer, along with your answer to the exercise. In the MyLab Math Interactive Assignment, click the **SHOW ANSWERS** button to check your answers. Correct any errors, or press the **PLAY** button for a video solution.

Multiple Choice. For Exercises 1 through 6 are Multiple Choice, choose the correct letter.

For Exercises 1 through 4, use the graph below.

Bushels of Oranges Picked

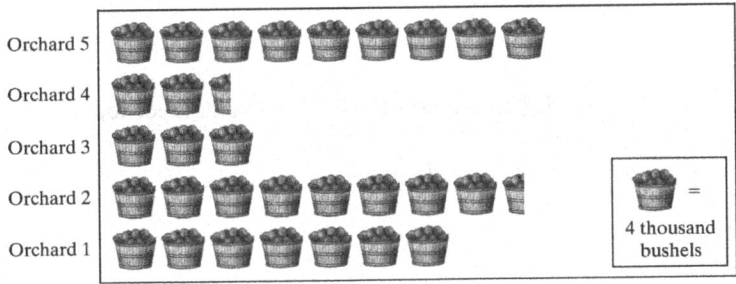

1. How many bushels of oranges did Orchard 1 produce?
   A. 7 bushels   B. 28 bushels   C. 28,000 bushels

2. Which orchard above produced the most bushels?
   A. Orchard 1   B. Orchard 2   C. Orchard 5   D. Orchards 2 and 5 produced the same.

3. How many bushels of oranges did Orchard 4 produce?
   A. 3 bushels   B. $2\frac{1}{2}$ bushels   C. 10 bushels   D. 10 thousand bushels

4. How many more bushels did Orchard 2 produce than Orchard 4?
   A. 24 bushels   B. $6\frac{1}{2}$ bushels   C. 24,000 bushels   D. 6000 bushels

5. Which square root is between the numbers 8 and 10?

    A. $\sqrt{36}$ B. $\sqrt{25}$ C. $\sqrt{49}$ D. $\sqrt{81}$

6. Which square root is between the numbers 6 and 7?

    A. $\sqrt{10}$ B. $\sqrt{13}$ C. $\sqrt{50}$ D. $\sqrt{40}$

True or False. *Choose the correct letter for Exercises 7 through 10.*

   A. True   or   B. False

7. The Pythagorean theorem applies to right triangles only.

8. A right triangle can have two 90° angles.

9. A right triangle has 3 sides, 1 side is called a leg and the other 2 sides are each called a hypotenuse.

10. The hypotenuse of a right triangle is the longest side.

Multiple Choice. Exercises 11 through 16 are Multiple Choice. Choose the correct letter.

11. Choose the degrees in the unknown section of this circle graph.

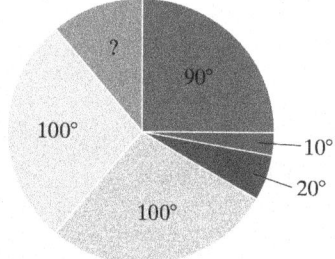

    A. 60°   B. 50°   C. 40°   D. 80°

12. The five colored marbles are placed in a bag. What is the probability of choosing a red marble?

    Red   Blue   Red   Blue   Red

    A. $\frac{1}{5}$ B. $\frac{2}{5}$ C. $\frac{3}{5}$ D. $\frac{4}{5}$

13. For the marbles in Exercise 12, what is the probability of choosing a green marble?

    A. 0   B. 1   C. $\frac{2}{5}$ D. $\frac{3}{5}$

For Exercises 14 through 16, use the data sets and choice below to answer.

    A. data set: 10, 10, 10, 10, 10   B. data set: 6, 8, 10, 12, 14

    C. data set: 8, 9, 10, 11, 12   D. they are the same

14. Which data set has the greatest range?

15. Which data set has the greatest median?

16. Which data set has the greatest mean?

Matching. Choose **two** data sets (two letters) in the right column that make each exercise statement true. Data sets may be used more than once or not at all.

17. equal means   A. 10, 20, 30, 30, 40, 50

18. equal number of modes   B. 9, 11, 11, 30, 48

19. equal ranges   C. 20, 25, 30, 35, 40

20. equal medians   D. 50, 55, 60, 65, 70

Review Exercises

In the **MyLab Math, Interactive Assignment, Review Exercises** section, there are algorithmically generated "Your Turn" exercises so that you can check your knowledge of some core concepts in this chapter. Insert a few sheets of paper in your Interactive Organizer to "record and show your work" along with the final answer.

Practice Chapter Test

- These exercises will help you practice for your chapter test.

General Directions: For each exercise, "show your work" by writing each step in the solution process within your Interactive Organizer, including your final answer. In the MyLab Math Interactive Assignment, click the Show Answer button to check your answer. Correct any errors, or press the **PLAY** button for a video solution.

The following pictograph shows the money collected each week from a wrapping paper fundraiser. Use this graph to answer Exercises 1 through 3.

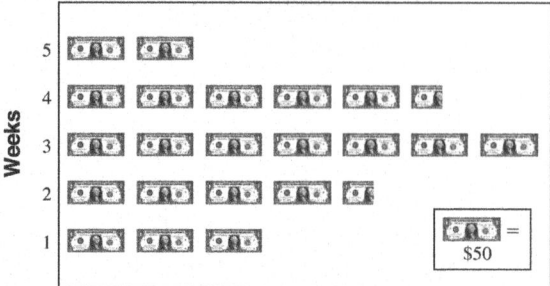

Weekly Wrapping Paper Sales

1. How much money was collected during the second week?

2. During which week was the most money collected? How much money was collected during that week?

3. What was the total amount of money collected for the fundraiser?

The bar graph shows the normal monthly precipitation in centimeters for Chicago, Illinois. Use this graph to answer Exercises 4 through 6.

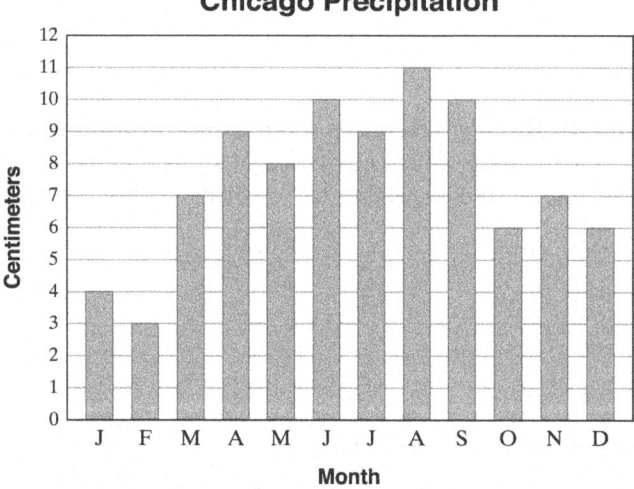

Chicago Precipitation

Source: U.S. National Oceanic and Atmospheric Administration, *Climatography of the United States*, No. 81

4. During which month(s) does Chicago normally have more than 9 centimeters of precipitation?

5. During which month does Chicago normally have the least amount of precipitation? How much precipitation occurs during that month?

6. During which month(s) does 7 centimeters of precipitation normally occur?

7. Use the information in the table to draw a bar graph. Clearly label each bar.

Most Common Blood Types	
Blood Type	% of Population with This Blood Type
O+	38%
A+	34%
B+	9%
O−	7%
A−	6%
AB+	3%
B−	2%
AB−	1%

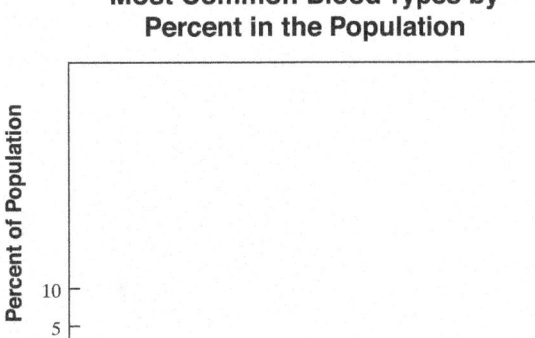

The following bar graph shows the annual inflation rate in the United States for the years 2008–2017. Use this graph to answer Exercises 8 through 10.

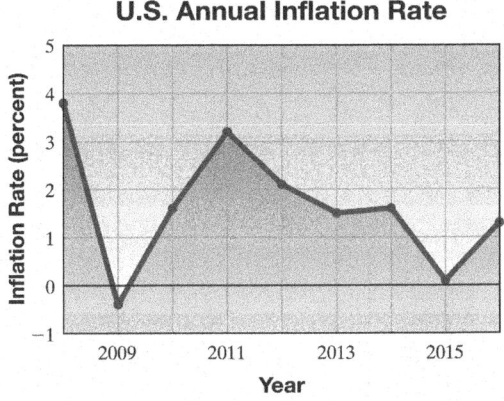

Source: Bureau of Labor Statistics

8. Approximate the annual inflation rate in 2014.

9. During which of the years shown was the inflation rate greater than 3%?

10. During which sets of years was the inflation rate decreasing?

The results of a survey of 200 people is shown in the following circle graph. Each person was asked to tell his or her favorite type of music. Use this graph to answer Exercises 11 and 12.

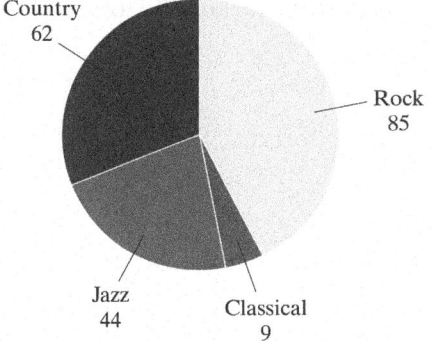

11. Find the ratio of those who prefer rock music to the total number surveyed.

12. Find the ratio of those who prefer country music to those who prefer jazz.

175

The following circle graph shows the projected age distribution of the population of the United States in 2020. There are projected to be 335 million people in the United States in 2020. Use the graph to find how many people are expected to be in the age groups given.

U.S. Population in 2020 by Age Groups

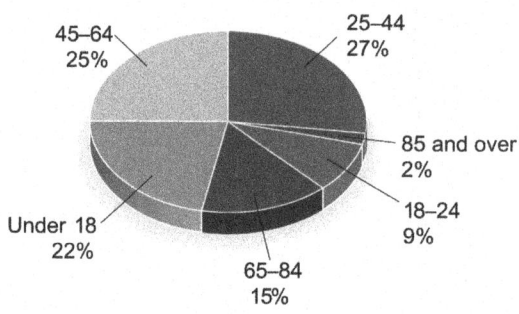

Source: U.S. Census Bureau

13. Under 18 (Round to the nearest whole million.)

14. 25–44 (Round to the nearest whole million.)

A professor measures the heights of the students in her class. The results are shown in the following histogram. Use this histogram to answer Exercises 15 and 16.

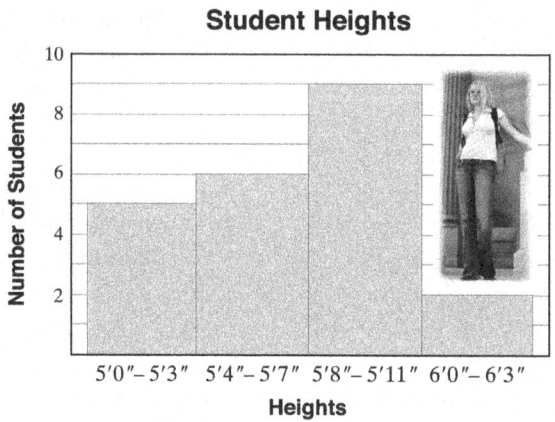

15. How many students are 5'8"-5'11" tall?

16. How many students are 5'7" or shorter?

17. The history test scores of 25 students are shown below. Use these scores to complete the frequency distribution table.

70 86 81 65 92

43 72 85 69 97

82 51 75 50 68

88 83 85 77 99

77 63 59 84 90

Class Intervals (Scores)	Tally	Class Frequency (Number of Students)
40 – 49		
50 – 59		
60 – 69		
70 – 79		
80 – 89		
90 – 99		

18. Use the results of Exercise 17 to draw a histogram.

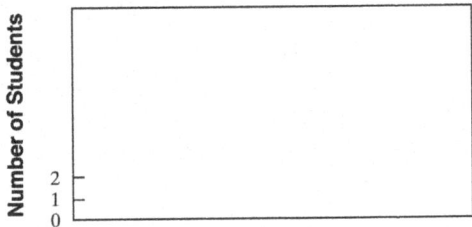

19. Find the range of the data items:
10, 18, 13, 16, 13

20. Use the data for Exercise 17 and find the range.

21. Use the frequency distribution table to find the following. Round the mean to 1 decimal place.

    a. mean   b. median

    c. mode   d. range

Data Item	Frequency
90	3
91	1
92	2
93	8
94	8

22. Use the graph of the data items to find the following. Round the mean to 1 decimal place.

    a. mean   b. median

    c. mode   d. range

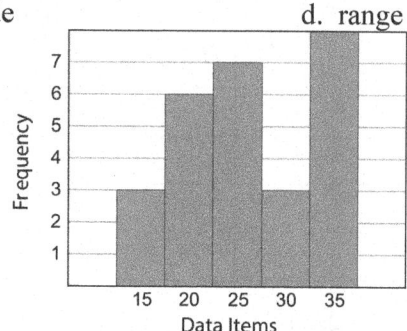

Find each square root and simplify. Round to the nearest thousandth if necessary.

23. $\sqrt{49}$ 24. $\sqrt{157}$ 25. $\sqrt{\dfrac{64}{100}}$

Solve.

26. Approximate to the nearest hundredth of a centimeter the length of the missing side of a right triangle with legs of 4 centimeters each.

27. Given that the following triangles are similar, find the unknown length, *n*.

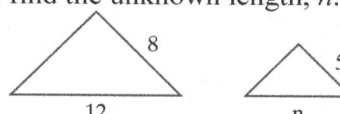

28. Tamara Watford, surveyor, needs to estimate the height of a tower. She estimates the length of her shadow to be 4 feet long and the length of the tower's shadow to be 48 feet long. Find the height of the tower if she is $5\dfrac{3}{4}$ feet tall.

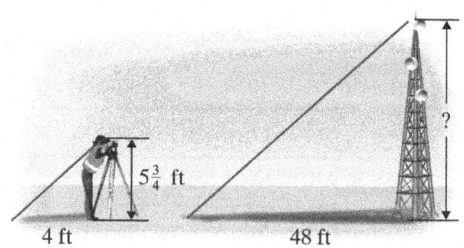

29. Draw a tree diagram for the experiment of spinning the spinner twice. Then use the diagram to determine the number of outcomes.

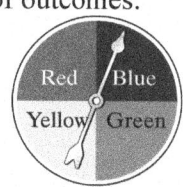

178

30. Draw a tree diagram for the experiment of tossing a coin twice. Then use the diagram to determine the number of outcomes.

Suppose that the numbers 1 through 10 are each written on same-size pieces of paper and placed in a bag. You then select one piece of paper from the bag.

31. What is the probability of choosing a 6 from the bag?

32. What is the probability of choose a 3 or a 4 from the bag?

Section 8.1 Lines and Angles

Objectives
A Identify Lines, Line Segments, Rays, and Angles
B Classify Angles as Acute, Right, Obtuse, or Straight
C Identify Complementary and Supplementary Angles
D Find Measures of Angles

Directions: Complete your Interactive Organizer by filling in the blanks and solving exercises as you complete each screen of the Interactive Assignment.

- For **WORK WITH ME** exercises, follow along and write each step needed and shown to solve, including the final answer.
- For **YOUR TURN** exercises, write the exercise generated for you in MyLab Math, then "show your work" by writing each step needed to solve, including the final answer.

Objective A: Identify Lines, Line Segments, Rays, and Angles

Watch the objective video.

_____ extends indefinitely in all directions.

A _____ is a flat surface that extends indefinitely.

A _____ has no length, no width, and no height, but it does have location.

A _____ is a set of points extending indefinitely in two directions.

How can you designate a line?

A _____ is part of a line with an end point.

How can you designate a ray?

Two rays with a common endpoint make up an _____ .

How can you designate an angle?

The common endpoint is called the _____ .

Helpful Hint
Naming an Angle
When there is no confusion as to what angle is being named, you may use the vertex alone.

Naming the Angle--
Using the Vertex Alone

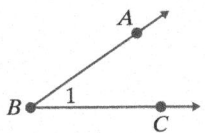

Name of ∠B is all right.
There is no confusion. ∠B means ∠1.

Naming the Angle--
Without Using the Vertex Alone

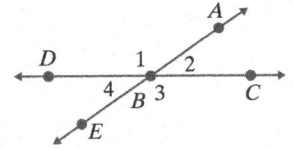

Name of ∠B is *not* all right.
There is confusion. Does ∠B mean
∠1, ∠2, ∠3, or ∠4?

YOUR TURN #1:   **YOUR TURN #2:**

YOUR TURN #3:   **YOUR TURN #4:**

Objective B: Classify Angles as Acute, Right, Obtuse, or Straight

Watch the objective video.

An angle can be measured in _____ .

How many degrees in a full revolution?   How many degrees in a half revolution?

An angle that measures 180° is called a _____ angle.

How many degrees is a quarter of a revolution?

An angle that measures 90° is called a _____ angle.

An _____ angle measures between 0° and 90°.

An _____ angle measures between 90° and 180°.

***VIDEO* WORK WITH ME.**

Classify each angle as straight, right, acute, or obtuse.

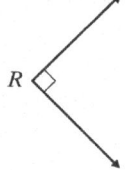

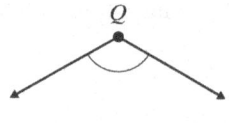

YOUR TURN #1:   **YOUR TURN #2:**

Objective C: Identify Complementary and Supplementary Angles

Watch the objective video.

Two angles that have a sum of 90° are called _____ angles. We say that each angle is the _____ of the other.

Two angles that have a sum of 180° are called _____ angles. We say that each angle is the _____ of the other.

***VIDEO* WORK WITH ME.**

Find the complement of a 23° angle.   Find the supplement of a 17° angle.

YOUR TURN #1:   **YOUR TURN #2:**

YOUR TURN #3:

Objective D: Find Measures of Angles

WORK WITH ME #1.
Find the measure of the given angle in the figure.

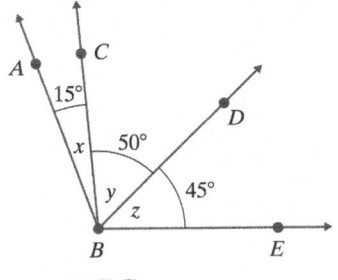

a. $\angle ABC$

b. $\angle DBA$

YOUR TURN #1:   **YOUR TURN #2:**

Two lines in a plane can be either parallel or intersecting. _____ lines never meet. _____ lines meet at a point.

Parallel Lines	Intersecting Lines

p ←————————————→

q ←————————————→

Parallel lines

Intersecting Lines figure with point T, lines l and k.

Intersecting lines

The symbol || is used to indicate "is _____ to." Here $p \parallel q$.

Some intersecting lines are perpendicular. Two lines are _____ if they form right angles when they intersect.

Intersecting Lines That Are Perpendicular

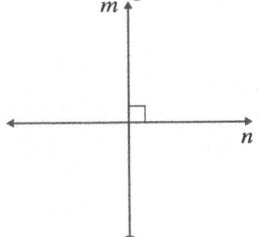

Perpendicular lines

The symbol ⊥ is used to indicate "is _____ to." Here $n \perp m$.

WORK WITH ME #2.

Two angles that are opposite each other are called _____ angles.

Vertical angles have the _____ measure.

Two angles that share a common side are called _____ angles.

Adjacent angles formed by intersecting lines are _____ . That is, they have a sum of _____ .

WORK WITH ME #3.

Find the measure of angles *x*, *y*, and *z* in the figure.

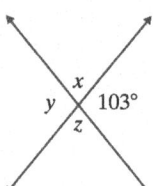

A line that intersects two or more lines at different points is called a _____ .

Line *l* is a transversal that intersects lines *m* and *n*.

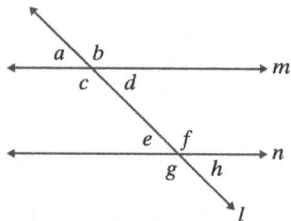

The eight angles formed have special names.

Corresponding angles:

Alternate interior angles:

When two lines cut by a transversal are *parallel*, the following statement is true:

Parallel Lines Cut by a Transversal
If two parallel lines are cut by a transversal, then the measures of _____ angles are
_____ and the measures of the _____ angles are _____ .

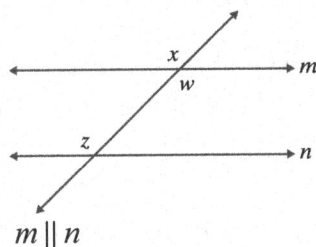

$m \parallel n$

Corresponding Angles:

Alternate Interior Angles:

WORK WITH ME #4.
Find the measures of angles *x*, *y*, and *z* in the figure.

$m \parallel n$

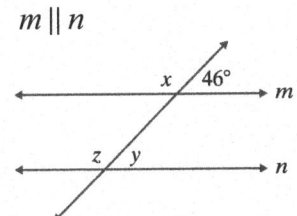

YOUR TURN #3:   **YOUR TURN #4:**

Section 8.2 Perimeter

Objectives
    A Use Formulas to Find Perimeters
    B Use Formulas to Find Circumference

Directions: Complete your Interactive Organizer by filling in the blanks and solving exercises as you complete each screen of the Interactive Assignment.
- For **WORK WITH ME** exercises, follow along and write each step needed and shown to solve, including the final answer.
- For **YOUR TURN** exercises, write the exercise generated for you in MyLab Math, then "show your work" by writing each step needed to solve, including the final answer.

Objective A: Use Formulas to Find Perimeters

Recall that the perimeter of a polygon is the _____ around the polygon. The perimeters of some special figures form patterns, which we call _____.

Perimeter of a Rectangle
    Perimeter =

In symbols, this can be written as

length

width   width

length

Perimeter of a Square
    Perimeter =
                =

In symbols,

side

side   side

side

Perimeter of a Triangle
    Perimeter =

In symbols,

side *a*   side *b*

side *c*

Perimeter of a Polygon

The perimeter of a polygon is the sum of the lengths of its sides.

186

WORK WITH ME #1.
Find the perimeter of each figure.

a.

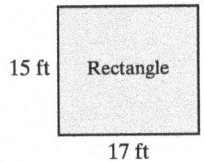

15 ft   Rectangle

17 ft

b.

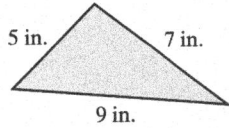

5 in.   7 in.

9 in.

c.

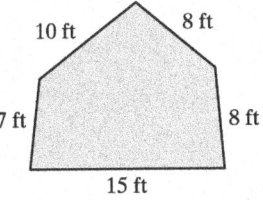

10 ft   8 ft

7 ft   8 ft

15 ft

d.

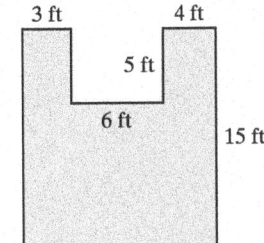

3 ft   4 ft

5 ft

6 ft

15 ft

YOUR TURN #1:

YOUR TURN #2:

YOUR TURN #3:

YOUR TURN #4:

Objective B: Use Formulas to Find Circumference

Watch the objective video.

Circumference of a Circle

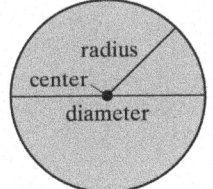

radius

center

diameter

Circumference

In symbols,

where $\pi \approx 3.14$ or $\pi \approx \dfrac{22}{7}$

VIDEO **WORK WITH ME.**

Find the circumference of

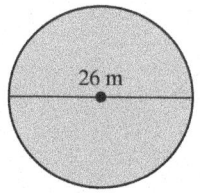

26 m

YOUR TURN #1:   **YOUR TURN #2:**

Section 8.3 Area, Volume, and Surface Area

Objectives
A Find the Area of Plane Regions
B Find the Volume and Surface Area of Solids

Directions: Complete your Interactive Organizer by filling in the blanks and solving exercises as you complete each screen of the Interactive Assignment.

- For **WORK WITH ME** exercises, follow along and write each step needed and shown to solve, including the final answer.
- For **YOUR TURN** exercises, write the exercise generated for you in MyLab Math, then "show your work" by writing each step needed to solve, including the final answer.

Objective A: Find the Area of Plane Regions

Recall that area measures the number of _____ units that cover the _____ of a plane region; that is, a region that lies in a plane.

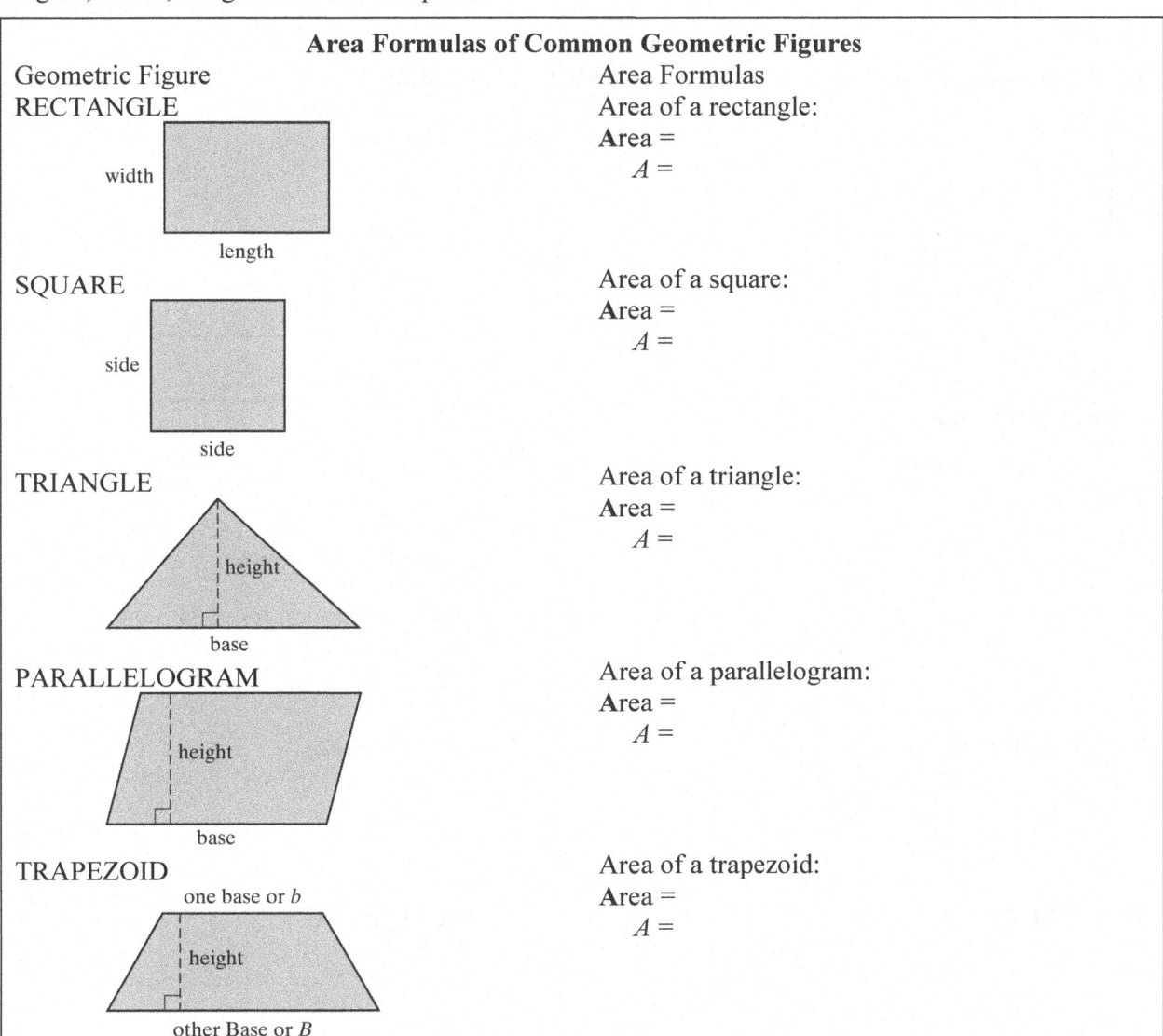

Area Formulas of Common Geometric Figures

Geometric Figure   Area Formulas

RECTANGLE   Area of a rectangle:
                                                        Area =
                                                        $A =$

SQUARE   Area of a square:
                                                        Area =
                                                        $A =$

TRIANGLE   Area of a triangle:
                                                        Area =
                                                        $A =$

PARALLELOGRAM   Area of a parallelogram:
                                                        Area =
                                                        $A =$

TRAPEZOID   Area of a trapezoid:
                                                        Area =
                                                        $A =$

WORK WITH ME #1.
Find the area of each geometric figure.

a.

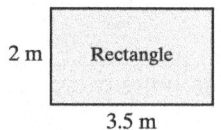

2 m   Rectangle

3.5 m

b.

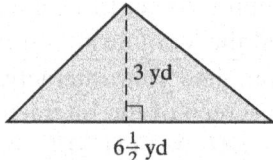

3 yd

$6\frac{1}{2}$ yd

WORK WITH ME #2.

Helpful Hint
The figure in the last example could also be split into two other rectangles.

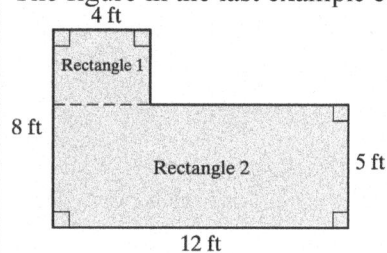

4 ft

Rectangle 1

8 ft

Rectangle 2   5 ft

12 ft

Area Formula of a Circle
Circle

radius

Area of a circle
Area =
$A =$

(A fractional approximation for π is $\frac{22}{7}$.)

(A decimal approximation for π is 3.14.)

YOUR TURN #1:   **YOUR TURN #2:**

YOUR TURN #3:   **YOUR TURN #4:**

Objective B: Find the Volume and Surface Area of Solids

We will find the volume and surface area of special types of _____ solids called _____ .
A solid formed by the intersection of a finite number of planes is called a _____ .

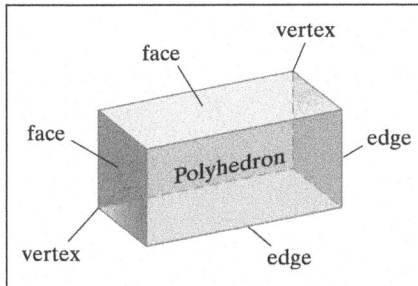

Each of the plane regions of a polyhedron is called a _____ of a polyhedron.

If the intersection of two faces is a line segment, this line segment is an _____
of the polyhedron.

The intersections of the edges are the _____ of the polyhedron.

_____ is a measure of the space of a region. The volume of a polyhedron, for example, is the
amount of space _____ .

The _____ area of a polyhedron is the sum of the areas of the _____ of the polyhedron.

Example of Volume
Polyhedrons

Actual size

1 inch

1 inch

1 inch

1 inch

Actual size

1 cm

1 cm

1 cm

1 cubic centimeter   **1 cubic inch**

The volume of a solid is the number of _____ units in the solid. A cubic _____ and a
cubic _____ are illustrated.

191

Example of Surface Area
Polyhedrons

Actual size

Actual size

1 cm

1 cm

1 cm

1 inch

1 inch

1 inch

Each _____ of the cube (above left) has an area of 1 square centimeter. Since there are _____ faces of the cube, the _____ of the areas of the faces is 6 square centimeters. Surface area can be used to describe the amount of material needed to _____ a solid. Surface area is measured in _____ units.

Volume and Surface Area Formulas of Common Solids	
Solid	Formulas
RECTANGULAR SOLID height width length	$V =$ $SA =$ where h = height, w = width, l = length
CUBE side side side	$V =$ $SA =$ where s = side
SPHERE radius	$V =$ $SA =$ where r = radius
CIRCULAR CYLINDER height radius	$V =$ $SA =$ where h = height, r = radius

CONE 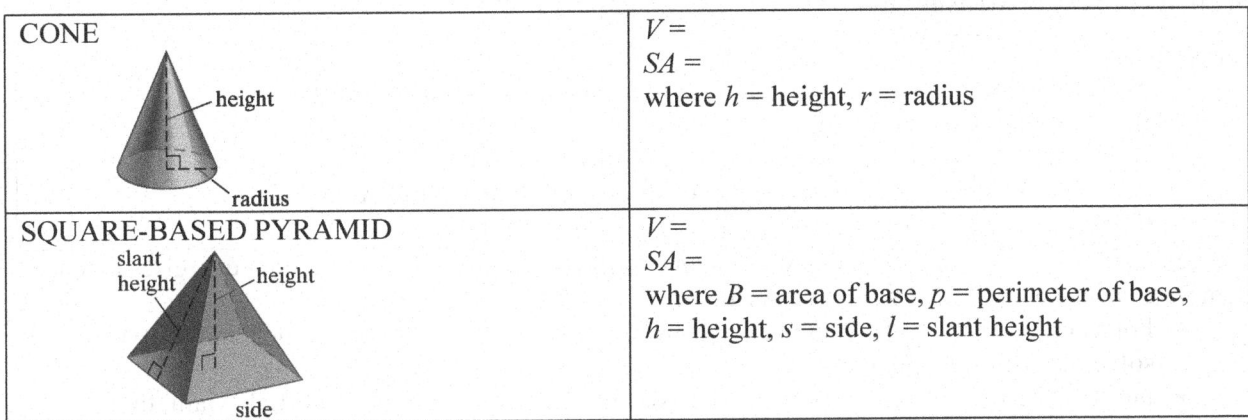	$V =$ $SA =$ where h = height, r = radius
SQUARE-BASED PYRAMID	$V =$ $SA =$ where B = area of base, p = perimeter of base, h = height, s = side, l = slant height

WORK WITH ME #1.

Find the volume and surface area of each solid. For formulas containing π, give an exact answer and

then approximate using $\dfrac{22}{7}$ for π.

a.

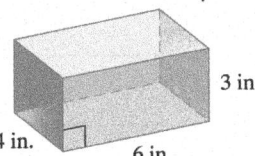

3 in.

4 in.   6 in.

b.

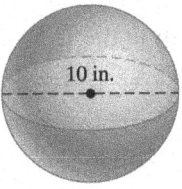

10 in.

c.   Find the volume only.

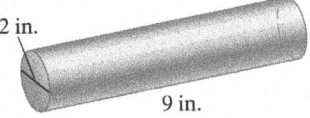

2 in.

9 in.

WORK WITH ME #2.

YOUR TURN #1:   **YOUR TURN #2:**

YOUR TURN #3:

193

Section 8.4 Linear Measurement

Objectives
> A Define U.S. Units of Length and Convert from One Unit to Another
> B Use Mixed Units of Length
> C Perform Arithmetic Operations on U.S. Units of Length
> D Define Metric Units of Length and Convert from One Unit to Another
> E Perform Arithmetic Operations on Metric Units of Length

Directions: Complete your Interactive Organizer by filling in the blanks and solving exercises as you complete each screen of the Interactive Assignment.

- For **WORK WITH ME** exercises, follow along and write each step needed and shown to solve, including the final answer.
- For **YOUR TURN** exercises, write the exercise generated for you in MyLab Math, then "show your work" by writing each step needed to solve, including the final answer.

Objective A: Define U.S. Units of Length and Convert from One Unit to Another

In the United States, _____ systems of measurement are commonly used. They are the United States (U.S.), or _____ , measurement system and the _____ system.

Common units of length in the U.S. measurement system are the _____, the _____, the _____, and the _____.

U.S. Units of Length
> 12 inches (in.) =
> 3 feet =
> 36 inches =
> 5280 feet =

To convert from one unit of length to another, we will use _____ fractions. We define a unit fraction to be a fraction that is _____ to 1.

Unit Fractions

$$\frac{12\ \text{in.}}{1\ \text{ft}} = 1 \ \text{ or } \ \frac{1\ \text{ft}}{12\ \text{in.}} = 1 \ \text{ (since 12 in. = 1 ft)}$$

$$\frac{3\ \text{ft}}{1\ \text{yd}} = 1 \ \text{ or } \ \frac{1\ \text{yd}}{3\ \text{ft}} = 1 \ \text{ (since 3 ft = 1 yd)}$$

$$\frac{5280\ \text{ft}}{1\ \text{mi}} = 1 \ \text{ or } \ \frac{1\ \text{mi}}{5280\ \text{ft}} = 1 \ \text{ (since 5280 ft = 1 mi)}$$

WORK WITH ME #1.
Convert each measurement as indicated.

a. 60 in. to feet

b. $8\frac{1}{2}$ ft to inches

c. 10 ft to yards

YOUR TURN #1:

YOUR TURN #2:

YOUR TURN #3:

Objective B: Use Mixed Units of Length

Watch the objective video.

VIDEO WORK WITH ME.

5 ft 2 in. = _____ in.

YOUR TURN #1:

YOUR TURN #2:

Objective C: Perform Arithmetic Operations on U.S. Units of Length

Watch the objective video.

VIDEO WORK WITH ME.

12 yd 2 ft
+ 9 yd 2 ft
_____

YOUR TURN #1:   **YOUR TURN #2:**

Objective D: Define Metric Units of Length and Convert from One Unit to Another

Watch the objective video.

The basic unit of length in the metric system is the _____ .

Metric Unit of Length
        1 **kilo**meter (km) = 1000 meters (m)
        1 **hecto**meter (hm) = 100 m
        1 **deka**meter (dam) = 10 m
              1 **meter (m)** = 1 m
        1 **deci**meter (dm) = 1/10 m or 0.1 m
        1 **centi**meter (cm) = 1/100 m or 0.01 m
        1 **milli**meter (mm) = 1/1000 m or 0.001 m

        km   hm   dam   m   dm   cm   mm

VIDEO WORK WITH ME.

1500 cm to meters   0.04 m to millimeters

YOUR TURN #1:   **YOUR TURN #2:**

Objective E: Perform Arithmetic Operations on Metric Units of Length

Watch the objective video.

VIDEO WORK WITH ME.

24.8 mm – 1.19 cm

YOUR TURN #1:   **YOUR TURN #2:**

Section 8.5 Weight and Mass

Objectives
A Define U.S. Units of Weight and Convert from One Unit to Another
B Perform Arithmetic Operations on U.S. Units of Weight
C Define Metric Units of Mass and Convert from One Unit to Another
D Perform Arithmetic Operations on Metric Units of Mass

Directions: Complete your Interactive Organizer by filling in the blanks and solving exercises as you complete each screen of the Interactive Assignment.

- For **WORK WITH ME** exercises, follow along and write each step needed and shown to solve, including the final answer.
- For **YOUR TURN** exercises, write the exercise generated for you in MyLab Math, then "show your work" by writing each step needed to solve, including the final answer.

Objective A: Define U.S. Units of Weight and Convert from One Unit to Another

Common units of weight (how heavy an object is) in the U.S. measurement system are the
_____ , the _____ , and the _____ .

  12 ounces   15 pounds   24 tons of garbage

U.S. Units of Weight
16 ounces (oz) = 1 pound (lb)
2000 pounds = 1 ton

To convert from one unit of weight to another, we will use _____ fractions.

Units Fractions
$\dfrac{16\ oz}{1\ lb} = \dfrac{1\ lb}{16\ oz} = 1$
$\dfrac{2000\ lb}{1\ ton} = \dfrac{1\ ton}{2000\ lb} = 1$

WORK WITH ME #1.
Convert as indicated.

a. 60 ounces to pounds

b. 4.9 tons to pounds

c. 89 oz = _____ lb _____ oz

WORK WITH ME #2.

YOUR TURN #1:   **YOUR TURN #2:**

YOUR TURN #3:

Objective B: Perform Arithmetic Operations on U.S. Units of Weight

Watch the objective video.

***VIDEO* WORK WITH ME.**

$$\begin{array}{r} 12 \text{ lb} \quad 4 \text{ oz} \\ -\ \ 3 \text{ lb} \quad 9 \text{ oz} \\ \hline \end{array}$$

YOUR TURN #1:   **YOUR TURN #2:**

Objective C: Define Metric Units of Mass and Convert from One Unit to Another

Watch the objective video.

The basic unit of mass in the metric system is the _____ .

Metric Unit of Mass

       1 kilogram (kg) = 1000 grams (g)
    1 hectogram (hg) = 100 g
    1 dekagram (dag) = 10 g
       1 gram (g) = 1 g
   1 decigram (dg) = 1/10 g or 0.1 g
  1 centigram (cg) = 1/100 g or 0.01 g
 1 milligram (mg) = 1/1000 g or 0.001 g

---

    kg hg dag g dg cg mg

VIDEO **WORK WITH ME.**

4 g to milligrams   6.3 g to kilograms

YOUR TURN #1:   **YOUR TURN #2:**

Objective D: Perform Arithmetic Operations on Metric Units of Mass

Watch the objective video.

VIDEO **WORK WITH ME.**

9 g – 7150 mg

YOUR TURN #1:   **YOUR TURN #2:**

YOUR TURN #3:

Section 8.6 Capacity

Objectives

        A Define U.S. Units of Capacity and Convert from One Unit to Another
        B Perform Arithmetic Operations on U.S. Units of Capacity
        C Define Metric Units of Capacity and Convert from One Unit to Another
        D Perform Arithmetic Operations on Metric Units of Capacity

Directions: Complete your Interactive Organizer by filling in the blanks and solving exercises as you complete each screen of the Interactive Assignment.

- For **WORK WITH ME** exercises, follow along and write each step needed and shown to solve, including the final answer.
- For **YOUR TURN** exercises, write the exercise generated for you in MyLab Math, then "show your work" by writing each step needed to solve, including the final answer.

Objective A: Define U.S. Units of Capacity and Convert from One Unit to Another

Units of _____ are generally used to measure liquids. Common units of _____ in the U.S. measurement system are the _____ ounce, the _____, the _____, the _____, and the _____.

U.S. Units of Capacity

        8 fluid ounces (fl oz) = 1 cup (c)
                    2 cups = 1 pint (pt)
                    2 pints = 1 quart (qt)
                4 quarts = 1 gallon (gal)

To convert from one unit of capacity to another, we will use _____ fractions.

WORK WITH ME #1.

Create a true unit fraction.

WORK WITH ME #2.

Convert each measurement as indicated.
a. 14 quarts to gallons   b. 42 cups to quarts

c. 58 qt = _____ gal _____ qt

YOUR TURN #1:   **YOUR TURN #2:**

YOUR TURN #3:

Objective B: Perform Arithmetic Operations on U.S. Units of Capacity

Watch the objective video.

***VIDEO* WORK WITH ME.**

$$\begin{array}{r} 3 \text{ gal} \\ - \ 1 \text{ gal} \ \ 3 \text{ qt} \\ \hline \end{array}$$

YOUR TURN #1:   **YOUR TURN #2:**

Objective C: Define Metric Units of Capacity and Convert from One Unit to Another

Watch the objective video.

The basic unit of capacity in the metric system is the _____.

Metric Unit of Capacity

        1 kiloliter (kl) = 1000 liters (L)
      1 hectoliter (hl) = 100 L
     1 dekaliter (dal) = 10 L
       1 liter (L) = 1 L
     1 deciliter (dl) =1/10 L or 0.1L
    1 centiliter (cl) = 1/100 L or 0.01 L
   1 milliliter (ml) = 1/1000 L or 0.001 L

---

    kl hl dal L dl cl ml

***VIDEO* WORK WITH ME.**

5600 ml to liters   0.16 kl to liters

YOUR TURN #1:   **YOUR TURN #2:**

YOUR TURN #3:

Objective D: Perform Arithmetic Operations on Metric Units of Capacity

Watch the objective video.

***VIDEO* WORK WITH ME.**

2700 ml + 1.8 L

YOUR TURN #1:   **YOUR TURN #2:**

203

Section 8.7 Temperature and Conversions Between the U.S. and Metric Systems

> **Objectives**
>   A Convert Between the U.S. and Metric Systems
>   B Convert Temperatures from Degrees Celsius to Degrees Fahrenheit
>   C Convert Temperatures from Degrees Fahrenheit to Degrees Celsius

Directions: Complete your Interactive Organizer by filling in the blanks and solving exercises as you complete each screen of the Interactive Assignment.

- For **WORK WITH ME** exercises, follow along and write each step needed and shown to solve, including the final answer.
- For **YOUR TURN** exercises, write the exercise generated for you in MyLab Math, then "show your work" by writing each step needed to solve, including the final answer.

Objective A: Convert Between the U.S. and Metric Systems

The Metric Act of _____ made use of the metric system legal (although not mandatory) in the United States.

You may be surprised at the number of _____ items we use that are already manufactured in metric units.

> **Examples of the Metric System in the U.S.**
>
>
>
>
>

Since the United States has not completely converted to the _____ system, we need to practice converting from one _____ to the other. There are _____ ways to perform these metric-to-U.S. conversions. We will do so by using unit _____ .

> **Length:**
>   Metric   U.S. System
>   1 m ≈ 1.09 yd
>   1 m ≈ 3.28 ft
>   1 km ≈ 0.62 mi
>   2.54 cm = 1 in.
>   0.30 m ≈ 1 ft
>   1.61 km ≈ 1 mi
>
>   1 yard
>   ▭▭▭▭▭▭▭▭▭▭▭▭▭▭
>   1 meter
>
> **Capacity:**
>   Metric   U.S. System
>   1 L ≈ 1.06 qt
>   1 L ≈ 0.26 gal
>   3.79 L ≈ 1 gal
>   0.95 L ≈ 1 qt
>   29.57 ml ≈ 1 fl oz
>
>   MILK   MILK
>   •1 qt   •1 L
>   1 quart   1 liter
>
> **Weight (mass):**
>   Metric   U.S. System
>   1 kg ≈ 2.20 lb
>   1 g ≈ 0.04 oz
>   0.45 kg ≈ 1 lb
>   28.35 g ≈ 1 oz
>
>   1 pound   1 kilogram

WORK WITH ME #1.

WORK WITH ME #2.

Convert as indicated. If necessary, round answers to the nearest two decimal places.

a. 86 inches to centimeters.

b. For an average adult, the weight of the right lung is greater than the weight of the left lung. If the right lung weighs 1.5 pounds and the left lung weighs 1.25 pounds, find the difference in grams. (*Source: Some Body!*)

WORK WITH ME #3.

YOUR TURN #1:   **YOUR TURN #2:**

Objective B: Convert Temperatures from Degrees Celsius to Degrees Fahrenheit

Watch the objective video.

| **Converting Celsius to Fahrenheit** |
| $F = \dfrac{9}{5}C + 32$ or $F = 1.8C + 32$ |

VIDEO **WORK WITH ME.**

A weather forecaster in Caracas predicts a high temperature of 27°C. Find this measurement in degrees Fahrenheit.

YOUR TURN #1:

Objective C: Convert Temperatures from Degrees Fahrenheit to Degrees Celsius

Watch the objective video.

Converting Fahrenheit to Celsius

$$C = \frac{5}{9}(F - 32)$$

***VIDEO* WORK WITH ME.**

77°F to degrees Celsius

Water freezes at _____

Water boils at _____

YOUR TURN #1:   **YOUR TURN #2:**

Chapter 8 Review and Practice

> Study Skills
> Chapter Vocabulary
> Getting Ready for the Test
> Review Exercises
> Practice Chapter Test

Study Skills

Directions: **Watch the Study Skills video.**

Chapter Vocabulary

WORK WITH ME.

Fill in each blank with one of the words or phrases listed below:

transversal	acute	obtuse	straight	adjacent	right	vertex	line segment

vertical	supplementary	ray	angle	line	complementary

1. A(n) _____ is a piece of a line with two endpoints.

2. Two angles that have a sum of 90° are called _____ angles.

3. A(n) _____ is a set of points extending indefinitely in two directions.

4. A(n) _____ is made up of two rays that share the same endpoint. The common endpoint is called the _____ .

5. A(n) _____ is part of a line with one endpoint. A ray extends indefinitely in one direction.

6. A line that intersects two or more lines at different points is called a(n) _____ .

7. An angle that measures 180° is called a(n) _____ angle.

8. When two lines intersect, four angles are formed. Each pair of angles that are opposite each other are called _____ angles.

9. Two of the angles from Exercise 8 that share a common side are called _____ angles.

10. An angle whose measure is between 90° and 180° is called a(n) _____ angle.

11. An angle that measures 90° is called a(n) _____ angle.

12. An angle whose measure is between 0° and 90° is called a(n) _____ angle.

13. Two angles that have a sum of 180° are called _____ angles.

207

Getting Ready for the Test.

- These exercises will help you avoid common errors while taking your chapter test.

General Directions: Read the exercise Write any notes or steps in this Interactive Organizer, along with your answer to the exercise. In the MyLab Math Interactive Assignment, click the **SHOW ANSWERS** button to check your answers. Correct any errors, or press the **PLAY** button for a video solution.

Matching. Match each word in the first column with its illustration in the second or third column.

1. line
2. line segment
3. ray
4. right angle
5. acute angle
6. obtuse angle

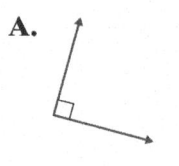

Multiple Choice. Exercises 7 through 16 are Multiple Choice. Choose the correct letter.

Use the given figure for Exercises 7 and 8.

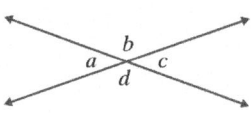

7. Choose two angles that have a sum of 180°.
     A. ∠*a* and ∠*c*   B. ∠*a* and ∠*d*   C. ∠*b* and ∠*d*

8. Choose two angles that have the same measure.
     A. ∠*a* and ∠*b*   B. ∠*c* and ∠*d*   C. ∠*b* and ∠*d*

Use the given figure for Exercises 9 and 10. For this figure, m ∥ n.

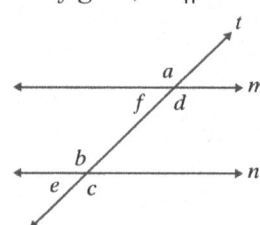

9. Choose two angles that have a sum of 180°.
     A. ∠*a* and ∠*e*   B. ∠*a* and ∠*d*   C. ∠*a* and ∠*b*

10. Choose two angles that have the same measure.
     A. ∠*a* and ∠*e*   B. ∠*a* and ∠*c*   C. ∠*a* and ∠*f*

For Exercises 11 through 16, the choices are below. Exercises 13 and 15 have two correct choices.

    A. perimeter   B. area   C. volume   D. circumference

11. Which calculation is measured in square units?

12. Which calculation is measured in cubic units?

13. Which calculation is measured in units?

*For Exercises 14 through 16, name the calculation (choices **A., B., C., D.** above) to be used to solve each exercise.*

14. The amount of material needed for a rectangular tablecloth.

15. The amount of trim needed to go around the edge of a tablecloth.

16. The amount of soil needed to fill a hole in the ground.

Review Exercises

In the **MyLab Math, Interactive Assignment, Review Exercises** section, there are algorithmically generated "Your Turn" exercises so that you can check your knowledge of some core concepts in this chapter. Insert a few sheets of paper in your Interactive Organizer to "record and show your work" along with the final answer.

Practice Chapter Test

- These exercises will help you practice for your chapter test.

General Directions: For each exercise, "show your work" by writing each step in the solution process within your Interactive Organizer, including your final answer. In the MyLab Math Interactive Assignment, click the Show Answer button to check your answer. Correct any errors, or press the **PLAY** button for a video solution.

1. Find the complement of a 78° angle.   2. Find the supplement of a 124° angle.

3. Find the measure of $\angle x$.

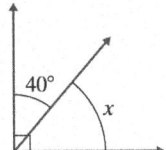

Find the measures of x, y, and z, in each figure.

4.

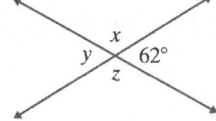

5.

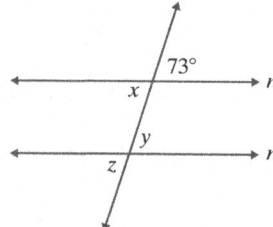

Find the unknown diameter or radius as indicated.

6.

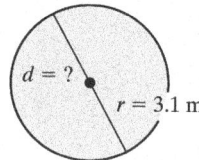

7.

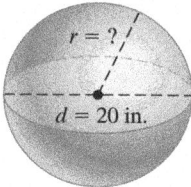

Find the perimeter (or circumference) and area of each figure. For the circle, give the exact value and an approximation using $\pi \approx 3.14$.

8.

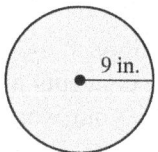

9.

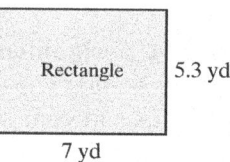

10.

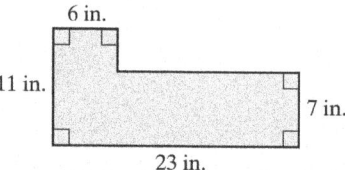

Find the volume of each solid. For the cylinder, use $\pi \approx \frac{22}{7}$.

11.

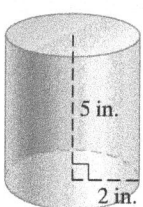

12.

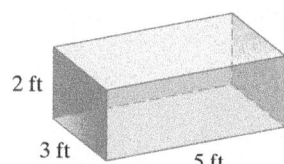

Solve.

13. Find the perimeter of a square photo with side length of 4 inches.

14. How much soil is needed to fill a rectangular hold 3 feet by 3 feet by 2 feet?

15. Find how much baseboard is needed to go around a rectangular room that measures 18 feet by 13 feet. If baseboard costs $1.87 per foot, also calculate the total cost needed for materials.

Convert.

16. 280 in. = _____ ft _____ in.

17. $2\frac{1}{2}$ gal to quarts

18. 30 oz to pounds

19. 2.8 tons to pounds

20. 38 pt to gallons

21. 40 mg to grams

22. 2.4 kg to grams

23. 3.6 cm to millimeters

24. 4.3 dg to grams

25. 0.83 L to milliliters

Perform each indicated operation.

26. 3 qt 1 pt + 2 qt 1 pt

27. 8 lb 6 oz − 4 lb 9 oz

28. 2 ft 9 in. × 3

29. 5 gal 2 qt ÷ 2

30. 8 cm − 14 mm

31. 1.8 km + 456 m

Convert. Round to the nearest tenth of a degree, if necessary.

32. 84F to degrees Celsius

33. 12.6C to degrees Fahrenheit

34. The sugar maples in front of Bette MacMillan's house are 8.4 meters tall. Because they interfere with the phone lines, the telephone company plans to remove the top third of the trees. How tall will the maples be after they are shortened?

35. A total of 15 gal 1 qt of oil has been removed from a 20-gallon drum. How much oil still remains in the container?

36. The engineer in charge of bridge construction said that the span of a certain bridge would be 88 m. But the actual construction required it to be 340 cm longer. Find the span of the bridge, in meters.

37. If 2 ft 9 in. of material is used to manufacture one scarf, how much material is needed for 6 scarves?